FLOWERS AND FRUITS
IN
THE WILDERNESS

FLOWERS AND FRUITS

IN

THE WILDERNESS

OR,

FORTY - SIX YEARS IN TEXAS AND TWO WINTERS IN HONDURAS.

Third Edition, Revised.

BY

Z. N. MORRELL,

AN OLD TEXAN,

"The desert shall rejoice, and blossom as the rose."—Isaiah.
"What thou seest. write in a book."—Rev.

Third Edition, Revised

ST. LOUIS

COMMERCIAL PRINTING COMPANY, 405 N. THIRD ST.

TO

R. E. B. BAYLOR and
HOSEA GARRETT,

MY EARLY ASSOCIATES AND CO-WORKERS, IN VIEW OF THEIR VALUABLE SERVICES TO SOCIETY AND THE CAUSE OF RELIGION

IN TEXAS,

AND AS A TESTIMONY OF MY CONTINUED CHRISTIAN CONFIDENCE AND AFFECTION,

This Humble Volume

IS

DEDICATED BY THE AUTHOR.

PREFACE.

I HAVE written this book because I thought I ought to write it. Some of the facts contained in it were known only to me. Many that were known only to me, and a few of my early associates in Texas, who are still living, would be lost when we are dead, and our time is very short. For the past fifteen years I have often been urged to write by those who were anxious to have some of the facts and incidents connected with the early society and the rise of religion in Texas preserved. Last October I was requested to write by the Baptist State Convention of Texas.

These friends and brethren did not ask me to write a book; but the work has grown upon me, as I toiled on, until it has assumed this form.

I have written under great personal affliction and physical suffering. Had I been able to go on with ministerial duties, the work would never have appeared. Hence I have written when I could do nothing else.

The work is written in a narrative style, because I could write in no other way. My memory served me better in this train.

My personal history in Texas, by following this plan, is interwoven with the state of society and the rise and progress of civilization and religion. If any imagine that I have intentionally made myself the hero of the book, through a spirit of selfishness and pride, let them remember that I am in my seventieth year, and conscious of the fact that "the silver cord" and "the golden bowl" will very soon be broken. My personal history is merely incidental, the remembrance of it serving to bring up the train of facts and incidents that I wished to record.

Some of the facts and anecdotes may appear silly and ludicrous; but the thoughtful reader will see that they are all illustrative of the state of things surrounding at the time.

If a humorous spirit is manifest, sometimes inducing a smile, be it remembered, that the sentiment expressed by Mr. Spurgeon, the great London preacher, in his preface to "John Ploughman," is in perfect accordance with my view, "that there is no particular virtue in being seriously unreadable."

My indebtedness is hereby acknowledged to Elder J. W. D. Creath, who has kindly allowed me the use of a large amount of material, which he has been years collecting, relative to the rise and progress of religion in Texas.

I am under special obligation to Elder M. V. Smith, for encouragement and assistance in many ways. Most of this work has been done under his roof. The offices of kindness shown me by this family have been many. He has aided me much in collecting statistics, arranging and revising. Without the assistance of brethren Creath and Smith I could not have succeeded.

Some may suppose that more statistics relative to the state of the country, men, and organizations, ought to have been given. The limits of the book would not admit of it. My purpose has simply been to lay the foundation for the historian. If I have succeeded in this, I have accomplished my most sanguine expectations.

Among some of the published documents in my possession are positive contradictions. In such cases I have adopted that which I believed to be correct. If *important* mistakes occur in the book, and proper testimony is privately given me, I will have corrections made in future editions.

I have not been able to get a single complete file of minutes.

If the reader of this humble volume shall have gained any useful information, or shall have been impressed with the power of the Christian religion in developing good society, and in furnishing rest and peace at the end of life, I shall be more than gratified.

Trusting in God, I send the little message forth, and ask a patient and thorough examination.

Z. N. Morrell.
Brenham, Texas, July 9, 1872.

PREFACE TO THE SECOND EDITION.

The first edition of one thousand copies of this work was issued a few months since. Its rapid sale, in the course of a few weeks, exhausted the entire edition.

A second edition, with some slight verbal corrections, mostly in names and dates, was put to press, but while in the process of printing, the plates and paper, with the office in Boston where the work was being executed, were destroyed in the recent extensive conflagration which visited that city.

Notwithstanding this serious loss, encouraged by the very flattering commendations and unusual sale of the work, it has been re-stereotyped, and a new improved edition now appears, which it is hoped may find increased favor in the eyes of the people.

Z. N. Morrell.
Brenham, Texas, January 25, 1873.

PREFACE TO THE THIRD EDITION.

For the past two years I have been urged to add to my former history; and though, as the reader has learned, I have met with many reverses in life, ending in having my stereotype plates burned, yet I comply with the demand, and earnestly ask my brethren and friends to aid me in the sale of this edition.

None who know me will expect me to write any more. Whether I live or die, arrangements are made to keep the book in circulation till some brother shall prepare a better history.

Z. N. Morrell.
Kyle, Hays County, Texas, June 14th, 1882.

CONTENTS.

CHAPTER I.

A TROUBLED PREACHER — IN 1835

CHAPTER II.

THE DECISION —1836.

CHAPTER III.

THE WILDERNERNESS SHALL BLOSSOM AS THE ROSE — 1836.

CHAPTER IV.

WAR—1836.

CHAPTER V.

AN EMERGENCY—1837.

CHAPTER VI.

LIGHT AHEAD — 1837

CHAPTER VII.

FIRST FRUITS IN THE MIDST OF TRIALS—1838.

PAGE

CHAPTER XII.

THE FIRST ASSOCIATION — 1840.

CHAPTER XIII.

THE DECISION UNDER DARK CLOUDS — 1842.

CHAPTER XIV.

MEXICAN INVASION — 1842.

CHAPTER XV.

WAR — 1842.

CHAPTER XVI.

AFFLICTION — 1843.

CHAPTER XVII.

DIMENSIONS AND TROUBLE IN EASTERN TEXAS — 1843.

CHAPTER XVIII.

ORGANIZATION IN MIDDLE TEXAS — 1844.

PAGE

CHAPTER XIX.

ANNEXATION AND EDUCATION — 1845.

CHAPTER XX.

PIONEER PREACHING — 1846.

CHAPTER XXI.

THE RACE TRACK AT SPRINGFIELD — 1846.

CHAPTER XXII.

TWO ASSOCIATIONS — 1847.

CHAPTER XXVII.

HARMONIOUS ACTIVITY — 1850 *TO* 1852.

CHAPTER XXVIII.

THE EASTERN CONVENTION — 1853.

CHAPTER XXIX.

REVIVALS — 1854 *AND* 1855.

CHAPTER XXX.

SKETCHES — 1857 *TO* 1867.

CHAPTER XXXI.

THE LAST TRIP OF A PIONEER — 1868.

CHAPTER XXXII.

THE CONCLUSION.

CHAPTER XXXIII.

AGE AND SUFFERING.

CHAPTER XXXIV.

A CONTRAST: 1872-1882

FLOWERS AND FRUITS.

CHAPTER I.

A TROUBLED PREACHER. — IN 1835.

DECEMBER, 1835 — October, 1871. Thirty-six years ago I planted my feet on the soil of Texas. How great the contrast in our circumstances then and now! And what a train of memorable events flashes upon my mind, as I review the conflicts and conquests of these years, — conflicts between barbarism and civilization, anarchy and well-regulated government; conquests of truth over error, and the faith of the gospel over priestcraft and superstition.

A brief history of some of the reasons that led the writer to Texas may be of interest. Some will, doubtless, suppose that a desire to roam over the earth, and to occupy fields hitherto uncultivated, was the only reason for this change. Not so. Perfect staisfaction was felt with the brethren, and the field of "good old Tennessee." But fourteen years of excessive ministerial labor in that State, attended with much exposure on long tours, swimming high waters, and preaching on an average about one sermon a day for nine years, brought on hemorrhage of the lungs, and made it necessary, according to the decision of two of our most eminent physicians, that I should emigrate south, and cease my ministerial labors, at least for a series of years. Language can never describe the sadness I felt, when I realized that the change must be made. Parting with fields, friends, brethren, and churches, where I had so long been accustomed to witness the triumphs of the gospel, to take up my abode among strangers, with shattered health, and with the responsibilities of my family and my ministry upon me, all this taxed my Christian fortitude to its utmost capacity.

My mind was turned to Texas in the fall of 1834. Its government was then very much disturbed. This obstacle in the way, and the additional fact that the iron arm of Catholicism was stretched over the whole land of Mexico, then embracing the State of Texas, did not make it a very desirable field for a Baptist preacher, who had always been accustomed to express himself boldly and independently. Catholicism, "the man of sin," I considered as a sworn enemy to me as a Baptist, and, after committing my way to God, I concluded to wait with patience for further light.

To move south was urged upon me as a present necessity, and in tears, with wife and four children, I wended my way south as far as Yallabusha County, Mississippi, and tarried to see what the gathering cloud in Texas would bring forth in 1835.

Sam. Houston was then in Texas among the Cherokee Indians, pulling the wires, by making friends with all the wild tribes of the red men of the forest; therby intending, with their aid, and with what emigration he could draw out from Tennessee and elsewhere, to set in motion "a little two-horse republic under the Lone Star," with the fond expectation that he would be its first president. This he had privately prophesied would be the case, in a confidential interview with his friend McIntosh, a deacon of the Baptist church at Nashville, about the time Houston abandoned the gubernatorial chair of Tennessee. Knowing that I had Texas on the brain, and in view of the fact that I had been a friend and admirer of Gen. Houston, for my encouragement, my good brother McIntosh committed this secret to me.

Months of gloom pass away, and I am still in Mississippi. Hemorrhage continues, and I wait the result of the climate upon my general health. With the prophecy previously alluded to in view, and with my mind turned south and west toward Texas, I was confidently expecting every day to learn that the political cloud had exploded, and that, as Sam. Houston's artillery would flash its lightning and roll its thunder across the prairies, the "Lone Star" colors would be raised, independence declared, the Republic formed, religious liberty established, and then I could go to Texas. But, alas! it is delayed.

With bleeding lungs, under special orders from my Tennessee physicians to refrain from preaching, and yet with fire burning in my bones to declare the way of salvation to the lost, I meet with brethren, acquaintances from Tennessee, and am urged to preach. The temptation was more than I could bear. By the way, I have always suffered from a weakness of this kind in the midst of gospel destitution. Three new churches are soon organized. All call me to serve as pastor; which is promptly declined, in consequence of my health and anticipated removal farther west, and I only consent to serve as a supply till other arrangements can be made. These churches, added to those already in existence in this section, make the organization of a new association, according to the decision of the brethren, a necessity. The time and place were agreed upon, and delegates appointed to meet at Troy, a little town on the Yallabusha River, in the fall of 1835. Having had a vast amount of trouble in the old State, above referred to, with the anti-missionaries, in organizing both churches and associations, and having been repeatedly ground by them in the flint-mills of tyranny and bigotry, in my feeble and worn condition, I confess I had no anxiety to engage in a similar contest in the new State of Mississippi. Notwithstanding my protest, I was selected as one of the delegates to bear a letter for the church. This I consented to do, after the church had passed a resolution, giving the delegates authority to withdraw the letter and return it, provided the association would not agree to an organization on gospel principles; allowing members to give to missions or not, as they saw proper.

These were days when the "old Simon pure" (as they called themselves) really thought it was a sin to belong to a Missionary Society, Bible or Tract Society, Sunday School, Temperance, Odd Fellows' or Masonic Lodge. All these were placed in the same category, and non-fellowship declared against every member of these institutions, whether Jew or Gentile. By the way, I had the audacity to contend for, belong to, and forward any or all of them, whenever opportunities were offered. We met on the ground at the appointed time and Elder Frank Baker was chosen as president, after a sermon by Z. N. Morrell. The president elect, be it remembered, was a decided anti-missionary, a man of wealth and great personal influence; but, withal, a warm-hearted, earnest Christian, never losing his equilibrium in the heat of debate. Courteous and kind in all his bearings, full of wit and humor, you could but love him, though you differed with him widely. After the temporary organization was complete a committee of five — the writer a member of it — was appointed to draft articles of faith, constitution and rules of decorum. This took place on Friday, Saturday was occupied with such business as could be properly attended to, waiting the report of the committee.

Preaching on Sunday, as was usual in that day, No cut and dried thirty-minute sermons were expected. We met at nine in the morning, and continued preaching until the sun was low in the west. The congregation on this occasion was very large, and preachers, one after another, occupied the entire day. All cool, doctrinal investigations.

Monday morning at length arrives, and the report of the committee appointed on Friday is called for. Now comes the tug of war.

The first objectional feature was found in an article making it criminal to contribute to the support of all the benevolent institutions of the day; making special declaration, that all members of the association should withdraw from Masonic Lodges, as they were nursed and fostered by the unfruitful *works of darkness*. As I was a member of the committee and the only minister it interfered with, I made my objections and enforced them; but finding myself overruled on the committee, I determined to let it pass, and fight it before the convention. If the reader will look over the long list of benevolent institutions, and remember that all these, without a single exception, were severely condemned by the article referred to, he will at once see a vast field opened for investigation. One by one they were taken up; objections made and answered. Finally, as the day was almost gone, the question was called for, and the objectionable feature expunged.

If my memory is not at fault, only one other article came up that demanded a decided protest. This article made all heretics accountable to the association. This was such a plain violation of the word of God, and looked so much like lording it over God's heritage, by wresting the power from the churches, which I then believed, and do yet believe, are the highest authority either civil or ecclesiastical under heaven; and in addition to this, considering my great exhaustion in consequence of the former debate, I resolved with the brethren who came with me to let it pass, and, by authority of the church of which I was a member, withdraw from the association and return home. The Author of the truth decided otherwise. Thanks to his great name, although I did not feel able mentally or physically to enter into an extended arument on this great question, yet I was able to cry *No* when the vote was taken, so as to be heard at the extreme limits of a large congregation. Profound silence reigned for some time. The president re-examined during the pause the letter we had borne, giving some discretionary power, and requested the association to call for the reasons of the long and emphatic *No.* By a vote of the body, after an almost unanimous *ay* in favor of the article, I was in my weakness called to the defense of the truth. With a church history in my hand, that was a favorite with me at the time, the minutes of some of the oldest associations of America, and with an open Bible, I took my stand, and gave the following reasons, which I will here condense as much as possible: —

1st. The law of God is violated in this article, by transferring to the association the responsibility touching heretics, that Christ and the apostles fixed upon the churches. If God in his word has referred matters of discipline to the church, as his executive, then it is positive rebellion against the King in Zion for any other body to assume disciplinary powers.

2d. A grand objection to the article is found in the fact, that the annual meetings of the association, to which you propose making heretics responsible, are nowhere in the Scriptures referred to as *annual* convocations. Baptists have ever held, that associations are bodies created by the churches and subject to their creative authority; simply as advisory councils, and to effect by combinations of churches what could not be accomplished by single churches in spreading the blessings of the gospel of peace.

3d. No power can create a power greater than itself. The association was here shown to be the creature of the churches, and therefore inferior in authority to the power that made it. Water cannot rise above its level. Man was not greater than his creator. No association, conference, or convention has the right to wrest a man from the authority of the church, and try him for heresy.

At the close of this discussion the convention adjourned, and repaired to the house of the president, one mile and a half distant, where preaching had been announced for the night. As before remarked, he was a man of wealth. Here we found a large house and a liberal heart. Preaching was soon over, and brother Baker, to-day our president, and to-night our landlord, with the brethren assembled in a large chamber, was delighting himself with good-humored boasts of belonging to the company of "old iron jackets and steel buttons," now and then making sarcastic hits at the missionaries and their folly in trying to do God's work. To this I could only reply, that we missionaries had one decided advantage. While the "iron jackets" boast of election and predestination, the missionaries are masters of the situation. "How is that?" cries the anti-missionary. We reply, "You worship a God that saw the end from the beginning, but left out all the means leading to and accomplishing the end. We worship an all-wise God who ordained the means leading to the end, as well as the end itself; and he has ordained the foolishness of preaching to save them that believe, as a means in his own hands." He asks, "Where do you find the word *means?*" 2 Samuel xiv. 14: "For we must needs die, and are as water spilt upon the ground, which cannot be gathered up again; neither doth God respect any person: yet doth he devise *means,* that his banished be not expelled from him."

I give this simply as a sample of the subjects and mode of discussion, during all the recesses, at our general convocations of that day.

Tuesday morning at eight o'clock the business of the body was resumed. Good attendance promptly at the hour. The objectionable article touching heretics was expunged, and the association was permanently organized, with brother Baker as moderator, and brother Wm. Minter clerk.

This about closed my labors in Mississippi, still looking at the political cloud hanging over Texas, until the sign should be given to pursue my journey south for my health.

On the first Sunday in December, 1835, after having administered both the ordinances of the church, baptism and the supper, — a cold, drizzly evening, — I returned home and found friends from Tennessee. Among them were two lawyers, Chester and Hayes, both nephews by marriage to General Jackson, then president of the United States; also brethren Moore and Hunt, deacons of the first two churches I ever organized; and last, but not least, my old physician, Dr. Butler. I had never been charged by my brethren and friends up to this time, nor have I since, with being over-prudent concerning my health. On this occasion, I received from my old physician a severe, but I suppose timely and merited, rebuke, for riding, preaching and baptizing on such a day. The doctor's language, as near as I can recall it, was as follows: "Hemorrhage of the lungs yet. You seem determined to kill yourself. I once thought you were a man of common sense, but I am now led to doubt it." An impression was made upon my mind, and I determined to *try* to do better in future. Some of the best preachers we have ever had in Texas have possessed zeal not well tempered with knowledge, and have been cut down in the prime of manhood. Let every preacher, and especially the younger ones, who read this, remember, that God has given his ministers a physical machinery that should be cared for, and that fearful responsibilities rest upon them in the midst of neglect.

Conversation now turned on Texas. These gentlemen were on their way to examine the country. The difficulties that had been in the way were by this time removed. Sam. Houston was at the head of a few hundred men, and calling on his friends in Tennessee and elsewhere for more; the war was in progress; troops were collecting at San Antonio; the independence of Texas was among the certainties of the future; would soon belong to the United States, and according to our opinion was the garden spot of the sunny South. These friends had just seen a letter from General Jackson to Adam Huntsman, of Tennessee, to the effect that Colonel Butler then minister plenipotentiary to Mexico, was instructed by the president to accede to propositions made by General Santa Anna, and close the treaty forth with for all the territory east of the Rio Grande. The supposition then was, based upon these facts, that a purchase would soon be made, and Texas would be free from Mexican rule and the tyranny of priestcraft.

The map of the country was placed before us, and the "Falls of the Brazos" was pointed out as a desirable locality for a colony to be formed. I was urged to join the party; the doctor underwrites for the effect of the climate upon my health; a night of anxious consultation with my family gets my consent, and, after two days are given to preparation, we are off for Texas.

CHAPTER II.

THE DECISION. — IN 1836.

"WHERE there is a will there is a way," is an old adage that is true, provided there is not too much undertaken and provided, further, that the enterprise is in accordance with the will of God. Two hundred and fifty miles lay between me and the nearest soil of Texas, and it was about five hundred to the "Falls of the Brazos." But few believed that I could make this journey on horseback, yet a still small voice from Macedonia was heard, and the angel of the vision bid me go forward, saying, "God hath much people in that land." We pass many things of interest; cross the father of American waters at Rodney, Red River at Alexandria, and reach old "Fort Gaines" on the Sabine, December 21, 1835.

Our company numbered six, and while crossing the Sabine, the ferryman related the following incident, that made a deep impression on our minds. The river at this crossing was the dividing line between Louisiana and Texas. Only a few days before a man rode up on the Louisiana side, evidently under great excitement, and at the top of his voice ordered the ferryman to bring over the boat. Supposing there was some emergency, the boat was promptly carried to the opposite shore, and the man landed as quick as possible on the Texas side. Just as he was ashore, an officer, with a body of men in pursuit of this refugee from justice, hailed on the eastern bank. The man, recognizing his pursuers, mounted his horse, rode up the hill entirely out of reach, and very deliberately made this short and pointed speech: "Gentlemen, I am just a little too fast for your sort. You have no authority out of the United States. I am entirely safe." Alighting from his horse and kissing the ground, he continued: "The Sabine River is a greater Saviour than Jesus Christ. He only saves men when they die from going to hell; but this river saves living men from prison."

We now began to realize the truth of what we had so often heard, that Texas was a place of refuge for scoundrels. Seeing that an impression was made upon our minds, and that all the company remained silent and thoughtful, the jocular ferryman, in order to dispel the cloud of gloom, continued: "And, gentlemen, what have you done that you have come to Texas?" The eyes of the lawyers, the doctor, the deacons, and the preacher, that composed our company, were turned inquiringly towards each other, and while each waited for another to reply, no one answered. After ferriage was paid, we steered our course towards San Augustine.

The conversation now turned upon the character of the inhabitants of the country through which we must pass, and a cloud of gloom for the evening hung over our spirits. We at no time would have felt disappointed at meeting robbers and cut-throats at any turn in the road. The evening's travel, through a poor, piney woods country, brought us to a house, a little more comfortable than we expected to find; but as there was a country store close by, at which we saw a company of suspicious-looking men just before darkness closed in upon us, we spent the first night in Texas in about the same fever of anxiety that we imagine was common with men at that day on their arrival in the State. The family treated us with marked civility, which of itself rendered much relief. Approaching San Augustine, the lands are more valuable, high, and greatly undulating.

On our arrival in the town we received reliable information of the battle, in which the Americans, on the tenth of December, 1835, took possession of San Antonio, after a hard struggle. Col. Milam was among the slain. We learned also that the soldiers were on their way home, after a three months' campaign.

Nacogdoches was the next town of importance before us. Regrets were expressed by the legal gentlemen that they were in Texas too late to share the honors of war, for which the battle of San Antonio had offered a good opportunity. This is no more than would have been expected from Hayes and Chester, now related by marriage to the "hero of New Orleans." Some of us, however, desired no such honors, especially as it would have required the peril of life, and in view of the responsibilities of wives and children behind us. High aspirations were at that early day freely indulged in by men who had acquired in some of the older States the smallest degree of distinction at the bar, or from some petty office. Such men easily worked themselves up to the belief that in this new country in a very short time they could become generals or statesmen among a people that they had been taught to believe consisted only of the dregs and renegades from the United States, north and south, as well as from various parts of Europe. Emigrants from Ireland and parts of the United States had settled in Texas as early as 1823; some even before that time.

It is true that just such an element was found in Texas at this time; thrown together in a wilderness country without restraint; mingling with the red men of the forest; roving over these wide-stretching prairies, overshadowed with a mild climate. Flowers bloomed and fruit was growing at almost all seasons of the year. Game of almost every description was so plentiful that it required but little more effort upon the part of man than the animal to obtain a subsistence. All these great blessings could only produce *one result* upon that class of the population above referred to, and that result was *indolence*, producting a retrograde movement on the intellect of the first immigrants, whose stock of knowledge, it was thought by the aspirants then entering the State, was of small importance at first. There were supposed to be no counterbalancing influences brought to bear. Here was a semi-savage, Mexican government, administered by a tyrant, himself under the tyranny of Catholicism, demoralizing in its character, and but one step in advance of the most degrading heathenism. With this picture before the minds of ambitious men, high anticipations were indulged in.

What position was offered to the writer of this humble narrative? He had but little of earth's goods, and had only acquired the name of an industrious, thorough-going "cane-brake Baptist preacher." Before his mind appeared at once a wide field for missionary operations, offering him honors far superior to any civil or military promotions. We could hear on inquiry of but one Baptist preacher in all this wide domain, and this was Daniel Parker, of "two-seed" notoriety, from Illinois, and whose writings had been scattered all over Tennessee. Here was a field large enough for several "cane-brake preachers" to labor in the organization of society on gospel principles. By this time my health was rapidly improving, and for several years I had no more hemorrhage.

We were soon in the old town of Nacogdoches, inquiring into the operations of this novel government, and the difficulties between Mexico and Texas, that had been instrumental in producing the then existing revolution. A long detail was given by the landlord and his associates of the grievances of the Texas family, — and a most remarkable family it was. One class has just been under consideration, composed, in Scripture phraseology, of "lewd fellows of the baser sort."

Among the company that relates to us the oppressions of the people, we come in contact with one who is the representative of another class. He is the proud, wealthy son of aristocracy. In youth he had been the child of affection, with all the advantages that wealth and education could confer. There were imprints stamped upon his deportment too delicate, too refined for us to mistake his mother. Noble blood coursed through his veins. Every flash of his eye, movement of his lips, nod of his head, and motion of his hand, with the beautifully rounded sentences and musical tones of his voice, declared to this auditors, while he repeats the grievances of this family of Texans, that he had passed through the universities of America, or some other clime; and it is written upon his demeanor that he came off with the first honors of his class. But, alas! there had been with him an evil hour. One evening, after his return from the schools, and while meditating upon the best plan to pursue in entering upon his vocation for life, he took tea at home, surrounded by all the sweet influences of a devoted mother and affecionate sisters, lighted a cigar, and took a stroll round town with a friend. A social glass for more pastime was indulged in. The first was exhilarating; a second was called for, and, under the excitement it produced, he forgot his promise to return home at ten, entered the hall of amusement, and unfortunately, lost money. Home was sought at a late hour, with spirits cast down. The next evening he determined to regain the lost treasure, and so continued for successive weeks. At last he decided that his money was lost by fraud, and, in a fit of desperation, slew his antagonist. Passion subsided; and, when reason resumed her sway, he saw that he had trampled upon the laws of his country, and must either submit to a terrible fate for flee from the State. During the lonely watches of the night he parted with all that he loved, and in haste crossed the Sabine. Crossing the *line*, he hurled back at his pursuers no such curses as those that escaped the former refugee, but resolved to be a sober man, and repented of his crime. But the laws of Tennessee demanded punishment instead of repentance, and he could not return.

He writes to his friends, from his home in the west, — tells of its fine climate, its beautiful prairies, its vast resources in the growth of cotton, corn and sugar, its immense herds of game, from the buffalo down to the mule-eared rabbit, and insists upon a visit from his friends. His letter is read by his old college associates; they come — return — report favorably, and families of wealth and enterprise gather around "our beloved Tom." Tom's mother was a noble spirit. This we could easily see from the bearing of the son. Other letters are written back by families brought to the State through his influence, all corroborating the statements of the unfortunate boy. In the language of David when his son was dead, the mother had for a long time inquired, "Can I bring him back again?"

David's reply to the question asked by himself was her reply: "I shall go to him, but he shall not return to me." Moved by this firm resolve, the parents, brothers, sisters, and a large company of the best people of the land had followed on.

Here you may see at a glance, that in 1835 there was a population, though small, of great variety. Among the "wood, hay and stubble" were found also "gold, silver and precious stones." Here was intellect of a higher type, and morality of a better cast, than we had formerly supposed; and among the people were a few whose "prayers and alms went up as a memorial before God."

Leaving Nacogdoches our legal associates expressed a decided anxiety to meet some of the soldiers, and get all the particulars of the battle. We had received from our landlord the names of several of our distinguished citizens from Tennessee, who had gone on this campaign.

Most of the lands for some distance were poor, except on the creeks, and a large quantity of this was at times overflowed. Prospects for health could not therefore be very flattering.

Approaching the Neches River, we met a company of soldiers, fresh from the seat of war. We agreed at once to get all the information from them that they were willing to communicate. Our horses were drinking in a beautiful stream as they rode up, and one of our number, appointed as spokesman, pleasantly and politely addressed them: "Good-morning, gentlemen. You seem to be travelling from the west, approaching the east. You must be in pursuit of light. We are from the east, approaching the west. We are in possession of some information, but would like to receive further news from the west. Suppose we alight, and, while our horses are grazing, exchange papers and notions for an hour.'' The foremost man, wearing a coon-skin cap, with the tail hanging down his back, his coat in rags, his shirt much the color of the hogwallow lands of the west, one leg of his pantaloons off at the knee, his feet protected by rawhide moccasins, with the hair on just as it grew on the cow, and resting in rawhide stirrups, readily agreed, and dismounting invited us all to make ourselves ''at home.'' We at once entered into an animated converstaion. The writer, in order to lead them out, and also to gratify the legal gentlemen from Tennessee, expressed some doubts as to whether the laws of nations would justify Texas in her revolt. In an instant the coon-skin cap was laid aside, the long black hair was thrown back from the forehead and eye that spoke intelligence of a high order. Legal questions too deep for a cane-break preacher were soon raised, and we soon ascertained that Tennesseeans could just about ask questions enought to keep Texans talking. Coon-skin caps, rawhide moccasins and stirrups and lariats, long hair and soiled garments, were not at all in their way. Having received the desired information, we parted with many hearty wishes for success in all future enterprises. They had clearly shown that Texas had sufficient reasons for revolution, and that there was ground for strong hopes of independence and anexation.

On parting some of the noble band took occasion, without speaking clearly, to indicate that they had seen initiations, passings and raisings among the honorable; and beyond had seen Mark Masters, Past Masters and Most Excellent Masters, and had passed under royal arches in their travels in the West.

We are now on our way, with the Neches River only a short distance ahead. Conversation of course turned on the soldier boys, on their way from San Antonio to their homes. One of the lawyers declared they were the smartest set of men, take them as a whole, he ever met. They numbered twenty-five or thirty, and while the man with the coon-skin cap was a leader in conversation, yet they all talked freely and intelligently. We had previously been told that we were ''green from the States;'' but now we began to realize it. Our appreciation of Texas intellect, you see, had thus far increased at every step. While we deplored the misfortunes that had brought refugees from justice to Texas, of which class ''our Tom'' was a fair specimen, we rejoiced that God had used the unfortunate boy as an instrument in his hands to bring to the State good men and pious women, with intellect, and wealth sufficient, under a favorable government, to develop the resources of the country, and form a nucleus around which good society might be formed. We frequently felt quite amused at the misapprehension under which people inthe older States labored, and sometimes soliloquized thus: ''He who goes in search of fools and greenhorns need not come to Texas.'' I felt some gratification, I freely confess, in seeing my legal associates outgeneralled upon questions of law by the soldiers; especially as I had been used as a cat's paw to bring on the conversation alluded to. Under these considerations and circumstances I began to feel, when only one hundred miles in the State, that warm admiration for Texas and Texans, which afterwards, in the strength of my manhood, burned in my soul, and which still lives to stimulate me in my declining years.

Neches River was soon crossed. We very soon observed some change in the country, occasionally a small prairie, as we approached the Trinity, at Robin's ferry. Crossing this stream, we entered, on rising the western bank, the first large prairie we ever saw, stretching further to the west than the eye could reach. It is hard to describe the emotions of the soul on looking for the first time across a vast prairie. The feeling is somehwat akin to that experienced on the first visit to the beach. Land and water there bore some resemblance. Here we took the old San Antonio road. But little timber was in sight during the evening, except a few post-oak groves; the trees so low that a Tennesseean was at a loss to know what was to be done for fencing. The land was superior, we thought, to what we had seen; but this did not fill the bill yet.

Night came on — a clear, cloudness night upon the prairie — who can describe it? No house in sight. In the midst of the grandeur of the scene, we confess to a feeling of loneliness. A few miles more, and a light in the distance gave evidence of the habitation of man. This consisted of one room, a small post-oak cabin sixteen feet square, dirt floor, and the habitation of man, wife and five children. Our company, it will be remembered, numbered six. Any port in a storm. I was appointed spokesman again. "Hem, halloo, can we stay all night?" Without asking any questions about our number, a voice was heard from the door: "We never turn anybody off, sir. Light, gentlemen, if you think you can put up with our fare." This invitation was given with such manifest hospitality that the party instantly dismounted, feeling quite at home. We were at once asked if we would have our horses *staked* or *hobbled.* Said I, "Staked"? for I had never heard the word used in this connection before. A lad about ten years of age was standing near the father, who addressed his mother in an under-tone, but within our hearing: "Perfectly green from the States." The preacher was sold again, to the great amusement of the company. After an explanation of the word *staked* was given, I decided that my mule would not stand comfortably at one end of a rawhide lariat, with the other tied to a post, and said "hobble." All the rest said "stake." Some of us slept inside the cabin, but more slept out. It was my time for amusement, when during the night the cry was heard, "Up and to your horses!" Some of the horses (for they were as "green from the States" as their masters) were tangled in their ropes, and whirling around like windmills; while others were down, kicking and cutting their legs terribly.

Twelve miles further, and we reached the Navasota River bottom — an ugly stream — land on the river all subject to overflow; must be very sickly, we thought. The timber consisted principally of hackberry, elm, oak, and some wild china.

We were now pressed hard by a former engagement. By agreement we were to meet the ex-congressman David Crockett, from Gibson County, Tennessee, on Christmas day, at the "Falls of the Brazos," and have a bear-hunt. I had lived in the same country with this distinguished person for eight years, and on several occasions had joined him in the chase. We were by this agreement bound to press forward with all possible haste, as it was now the 26th of December, and a long day's ride between us and the place. Some of our two-hundred-dollar celebrated horses were quite lame, in consequence of getting so much "green" peeled off at one time, during the previous night, and our progress by this accident was considerably retarded.

A hard ride brought us to brother Jesse Webb's, whom we had known from character in Tennessee, now living near the "Little Brazos," and near the present locality of Calvert, on the Houston and Texas Central Railroad. Here we were kindly received by a brother, whose eyes, about ten years after, it was made the writer's duty to close, when death had done his work.

While staking our horses to grass, he pointed to a post where a horse was standing tied a few nights before, when an Indian came, cut the rope, and carried the animal away. We had given our horses no corn since crossing the Trinity. Some of their legs were sore from "walking ropes;" we had travelled since morning near fifty miles; must go on the morrow through a country infested with Indians, — all of which led some of the company to plead right earnestly for corn. We were informed that the supply on hand was needed for bread; that the family had lived without bread, except one barrel of flour, while the present crop was being made; and that one party stood guard, while the other ploughed, during the entire season. Beside this, a widow hard by must be fed from the crib, or suffer.

The preacher could not be spokesman for corn after these statements; but some of the party offered ten dollars for a bushel, in view of the fact that we might have to make a race from the Indians before another sun went down. The corn was brought; not because ten dollars per bushel were offered, but because of genuine hospitality, as the sequel will show. Our supper consisted principally of *bear bacon, turnip greens,* and fresh *buffalo beef.* You may be sure that, tired and hungry as we were, we did ample justice to the feast around such a borad.

The two deacons of our company, who had served with Sam Houston in the Creek war, and had fought by his side at the time he received the wound that helped to make him governor of Tennessee, both felt the importance of a guard for our horses, and it was detailed.

The fact being ascertained that a preacher was to be entertained for the night,

God was thanked by our pious host, and upon invitation we surrounded our first family altar in Texas. The guard was changed, and we retired towards midnight to rest. It was rather a novel idea to some of us, that men could sleep under such circumstances. Daylight came; horses were all safe, and after a much relished breakfast, and horses prepared for the journey, our bill was called for, and a genuine Texas bill was presented: ''Call and see me on your return.'' Some of the crowd took care to pay the lad for the corn, and paid him well. A sharp lookout was kept all day for Indians. Game was seen in every direction, in great numbers. We saw land during the day that had been cultivated, without fence, and we supposed, from appearances, that it had grown fifty bushels to the acre. Some of the corn was not then gathered. At night we were at the ''Falls of the Brazos,'' the long-sought resting-place for a season.

Here we found but one family. There was, close by, the camp of about forty Tennesseeans. They were all out on Little River, hunting lands. Upon inquiry, we found that David Crockett had not arrived, and consequently the bearhunt with him was a disappointment. After a little rest, we began a thorough examination of the country for ourselves. Our expectation as to the great value of the lands was fully realized. The country was all we could desire, — lands very rich, range extraordinarily good, wood and water plenty, and the prospect for health very flattering. The river at this time was very low at this point, — not over knee-deep to our horses, — the falls about ten perpendicular feet, and the water below them abounding with fish. We examined the place minutely with reference to its capacity to run machinery. A few old Texans yet live who remembered that the stream was quite narrow at this point in 1836.

After a staisfactory examination was made of the surrounding country, we left for the three forks of Little River, about thirty-five miles south-west. No roads, except small trails through this wilderness. Great uneasiness was felt at this time relative to Indian depredations. There were fears of a general outbreak, predicated upon the amount of stealing going on through the country since the war began between the Americans and Mexicans. The Mexicans were evidently encouraging all the wild tribes to exterminate the colonists.

General Sam Houston now had use for all his ingenuity among the Indians to evade the fatal catastrophe. The war between Indians and colonist was also being hurried on by the land speculators, as their lands were valueless without an increase of population in this part of the State. So much for the state of the country.

Our trail was quite difficult to follow. Mustang, or wild Mexican horses, had trails going in almost every direction. Our path was distinguished from theirs by the prints of horse-shoes. Occasionally a buffalo trail crossed ours, but the difference was easily detected. Yonder in the distance was seen a large herd of mustangs, and on every hand great numbers of deer. We had become so much accustomed to the latter that by this time they attracted but little attention. All were watching for an Indian and a buffalo.

Rising the hill across Elm Creek, the leader of our party cried out, "What is that, with a hump on his back?" The animal, startled by the sight of man and the sound of the human voice, gave that noise peculiar to the buffalo tribe, and off went a large herd, making the earth almost tremble beneath the terrible stampede. Here was an open field, no timber, a chance for a fair race; not one of us had ever seen a buffalo in his life, and it was really amusing to see old men, farmers, and deacons in the Baptist church, with the representatives of three honorable professions, all forgetting that they were seven or eight hundred miles from home, with horses jaded, and some of them not well since their performance on the ropes, running their animals at the top of their speed, and shouting with all their might. Here we went, helter-skelter. Hurrah boys! bang, bang, bang, went guns and pistols, and away went the herd, following close upon the heels of their leader. They ran scientifically, with the right foot before, a side at a time, for three or four hundred yards. Then the leader would change and run with the left foot before, every buffalo following him making the same change. The writer, poor fellow, rode a mule, and it would show its blood. It would run with all its might toward the herd; but when it would get within forty yards, and sniff the peculiar odor that escapes the buffalo in the chase, it would invariably shy round. Whenever I would get near enough and ready to shoot, I would find my mule at right angles with my game, and bounding rapidly away. I thought to myself, "No meat for me, unless this part of the performance can be changed." The herd was soon gone, with no damage done that we could discover, and it was with difficulty that we found our way back to the trail. The supposition was that we had gone about two miles in this race. When night overtook us, we had travelled thirty miles, and were at the only house we had seen since starting in the morning.

Here on Little River we found the forty Tennessee landhunters, and among them Deacon Cartmell from the Baptist church in Nashville. What a joyful and unexpected meeting it was! "Why," said Brother Cartmell, "I never expected to see our cane-brake preacher again. I heard that you had died with hemorrhage of the lungs." My health, by this time, was almost entirely restored, and my voice clear and full; at least this was the decision of my comrades on our return to the trail, after the buffalo chase was over. Mr. Childress, whose wife was a Baptist, was the occupant and owner of that little lone cabin in this wilderness, and the family and land-hunters decided that they must have a sermon after supper; and accordingly I preached my first sermon in Texas, in camp, on the thirtieth of December, 1835. Here we spent a few days, and went out near where the city of Austin is now located, on the Colorado River.

After an absence of about twelve days Deacon Hunt and I were back at the

Falls; the other four having remained with the land-hunters. Our mission was accomplished, and we were seriously considering the propriety of moving the family eight hundred miles, to settle in the wilderness. The climate certainly would be suitable for my condition, and as I could neither preach in Tennessee nor Mississippi without endangering my life, I felt a strong inclination to make the change. After much prayer and meditation my mind was made up, and I thank God yet for the decision.

CHAPTER III.

THE WILDERNESS SHALL BLOSSOM AS THE ROSE. — 1836

NONE can tell, except a husband and father who has experienced the emotion, the feelings with which I mounted my animal, that I hoped would carry me to my loved ones in the east. On the tenth day of January, 1836, Deacon Hunt and the writer were "homeward bound." In the evening we were back at brother Webb's. Near the place where Wheelock is now located we found a garden full of vegetables, presenting more the appearance of spring than winter. I had just left wild rye and grass in the Brazos River bottom over knee high, and here the potato vines were not killed in the fields. While Texas has changed from a savage to a civilized state during these thirty-six years, its climate has, in many respects, undergone material changes.

January, 1836, wore the garments of spring at the Falls of the Brazos. I am now writing in Washington County, two degrees south; it is January, 1872, and, as I look through my window, the earth, the trees and houses are clad in garments of snow. It is much colder in winter now, more sultry in summer, and, as a rule, rains more now in one year than it did then in three.

Nothing of special importance occurred until we reached the old town of Nacogdoches in the east. Having an important engagement to be at a point beyond Nacogdoches on Sunday night, that must be met "if the Lord will," and being twenty miles west of the town when the sun of Sunday morning rose, I felt compelled to violate my former custom and travel on the Lord's day. My mind was by no means at ease. Several Sundays had come and gone while we were in the wilderness, and only one sermon had been preached, and that on an evening during the week. This was by no means the course I had pursued for fourteen years in Tennessee. My very soul burned within me to preach Jesus.

An election was in progress when I reached the town. This was the law and custom of the country in that day. Here was a large crowd of Americans, Mexicans, and Indians of several different tribes. My mule was soon tied, and after consultation with my great Master — for I had no one else to consult with — I decided to preach, and began looking around for a suitable place. Near by the vast crowd I saw the foundation timbers of a large framed building already laid. No floor had been laid, nor upright pieces raised. No sooner discovered than I selected one corner of this for a pulpit, — the sills and sleepers already laid and well adjusted would answer for seats. I held up my watch in my hand, and cried at the top of my voice, "O-yes! o-yes! o-yes! everybody that wants to buy, without money and without price, come this way," — and commenced singing the old battle-song: "Am I a soldier of the cross?" Before I finished my song there was around me a large crowd of all sorts and sizes and colors. A brief prayer was offered, and the two verses sung, "'Tis religion that can give," amidst profound silence. Astonishment, rather than reverence, was stamped upon their features. Across the street was a large upper gallery, and by this time it was full of ladies and gentlemen. Just at this point some wagons and a carriage, evidently belonging to movers, drove up close to where I was standing, and I recognized brother Wm. Whitaker and family, from Hardiman County, Tennessee, three of whose daughters I had baptized in the old State. The preacher who reads this will understand the effect this produced upon the speaker. My text was announced from Isaiah xxxv. 1: "The wilderness and the solitary place shall be glad for them; and the desert shall rejoice and blossom as the rose." Never did the cane-brake preacher receive better attention. God blessed me with great liberty for one hour, amid many tears shed all around me. The congregation was dismissed in due form, and there were many hearty shakes given the strange preacher's hand. My soul was full to overflowing, and at that moment I believed the text. God has not disappointed me.

We took a Red River steamer at Nachitoches for Natchez, and reshipping there landed at Memphis. Passing through the western district of Tennessee, I preached to my old churches, having been absent from them, amid very many trials, for about six months. My business was closed up as rapidly as possible; arriving in Mississippi, found my loved ones all well and willing to share with me the fortunes of Texas, be they good or bad.

CHAPTER IV.

WAR. — 1836

PREPARATIONS preceding a removal from an old country to a new one remind me very much of the preparation necessary to be made in going from time to eternity. A great many articles formerly of use must now be dispensed with, and other articles at a *great price* must be obtained. Preparation necessary being made, we were all soon on board the steamer "Statesman," about the first of April, 1836. After a smooth and quick voyage we were landed at Nachitoches, Louisiana; teams were purchased and provisions laid in for the long overland journey.

On the first day and every day till we reached the Sabine, we met families running away from Texas. On the second day of March the declaration of independence was signed by the convention in session at Washington, Texas, declaring Texas a *sovreign, free and independent Republic.* Exasperated by this bold stand of the people, and in view of the defiance of Travis, Crockett, and Bowie at San Antonio, the Alamo under the eye of Santa Anna had been surrounded on Sunday morning, March 6th, by the entire Mexican army, and one hundred and eighty-eight men brave men put to the sword, and the Texans were at this time in full retreat under Sam. Houston, and the Mexicans in full pursuit under Santa Anna. I was upbraided by everybody I met, and by some cursed as a fool, declaring that my family would be slain either by Mexicans or Indians before we would get far beyond the Sabine. Seldom in life had I turned back, and trusting in God, we travelled on.

Reliable information soon met us to the effect that General Houston, with his forces, consisting of seven hundred and eighty-three men, had engaged the Mexicans near two thousand strong, routed the army and captured Santa Anna. Mexican loss was six hundred and thirty killed, two hundred and eight wounded, and seven hundred and thirty prisoners, with all their camp equipage and the military chest, containing twelve thousand dollars. Texan loss was only eight killed, and twenty-five wounded. This famous battle, which turned the scales in favor of freedom, for a people borne down by oppressions grievous to endure, was fought on the twenty-first day of April, 1836.

Santa Anna, by permission of General Sam. Houston, sent a courier to his general next in rank to himself, ordering him and all the Mexican forces out of Texas. The cowardly Mexican tyrant now sat crouching at the feet of the "Hero of San Jacinto," no doubt dwelling upon the fate of the Alamo, and the murder of the one hundred and eighty-eight brave men by his own order, on Sunday, the sixth of March; and well remembering the order he had given to generals in command of the different divisions of his army, to shoot all the prisoners that fell into their hands, which had resulted in the coldest-blooded murder of the brave Fannin, and three hundred and thirty men at Goliad, on Sunday, March 27th, who were promised, if they would surrender, to be treated kindly as prisoners of war, and to be sent in vessels at once to the United States. Trembling for his own personal safety, and for the prisoners taken at the same battle, he was willing, at least for the time being, for peace.

The families were now all invited back. Some of them had taken such a fright that they did not return for a year; others never unloaded their wagons. We stopped at San Augustine several days, and rested the family and teams. While there I met with Sam. Houston for the first time in Texas, then suffering from the wound received at the battle of San Jacinto. He entertained no doubt of the success of the "little two-horse republic" that he had seen with prophetic eye years before, while yet in Tennessee.

The Cherokee Indians then occupied the territory north of Nacogdoches, and west to the Neches River. From them we purchased some cattle, and moved on as rapidly as possible to the Brazos. On our arrival everything looked lonesome and dreary. With depredations constantly going on in the west by the Mexicans, and on the north by the Indians, it required fortitude to stem the current. But God I believed had sent me and mine to Texas, and it would never do to run. The people made some corn, notwithstanding the runaway scrape, before alluded to. We had to move cautiously. We heard of continued threats of Mexican invasions. The Texas army remained in the field; but there was very soon not a dollar in the military chest. The year 1836, although flushed with victory at San Jacinto, was a year of great trial for the Texas people. Numbers of our men were absent from their families nearly the entire year, either in the Texas army, or west of the Trinity making a crop, while their families remained in the east. Indian depredations were now beginning in good earnest. Truly in every way it was a year of trial that the few living survivors have not forgotten.

While travelling with my family in the month of November, 1836, from the neighborhood of Wheelock to the Falls, a young man by the name of Reed being our only company, we camped near a house a little off from the main road. Guns were all put in order, and my two little sons, aged thirteen and eleven years, and young Reed, were ordered to lie down with their shot-bags round their necks and their gun-locks under their blankets to keep the powder dry. At that day we used flint and steel locks exclusively. About eleven o'clock at night, the Indians, about one hundred and fifty in number, approached our camp. Our faithful dogs raised the alarm, and on rising to our feet we discovered that the Indians were in close contact with the dogs. Every man and boy was ordered up, and with guns in hand, two men and two boys soon stood on the opposite side of the fire from the Indians. Campers will all see at once the advantage gained by this tactics. The Indians were frightened when they saw the guns, and through the fire we could see them skulking away into a ravine close by. Of course there was no more sleeping done that night in our camp. Morning came, and the God of Jacob was praised for our deliverance.

About eleven o'clock we were overtaken by a company of fifteen soldiers, most of them on foot, and on their way to the fort at the Falls. We gave them at noon all the provisions we had, for they were quite hungry. My little children were greatly distressed as they saw these hungry soldiers devouring the last of our provisions. My feelings I will not attempt to describe. It was my duty, and, trusting in God and these soldiers, we travelled on in the evening as cheerfully as we could. They agreed to camp with us at night. The Indians we then knew were below, and between us and the settlements. Near the little Brazos our teams were all hobbled and staked, a fire built, and soon the soldiers were scattered through the adjoining woods in search of game. Their commissary stores, and ours too, were exhausted. One gun after another was fired; the axe and water bucket had disappeared from our camp; the noise of the axe was soon heard; a tree fell; and several turkeys were brought in, and a bucket of honey. All were too hungry to pick turkeys; they were skinned, divided in two pieces, and hung on sticks before the fire. This was our first meal as a family without bread. Will it be the last? The news was soon received that Harvey and his family, occupying the house near our last camp, were all killed by the Indians in the evening after we left. These were, without doubt, the same Indians that approached our camp the night before. We felt exceedingly sad after hearing the fate of the unfortunate family, but doubly grateful to God in consequence of our escape on the previous night.

These soldiers, with whom we had divided bread in the journey, and who, after eating bountifully of turkey and honey, sat cheerfully round the camp fire, treated us with marked courtesy and kindness. A majority of them well knew how to do it, because of their training in youth. Most of these were young men, who had come to Texas as forerunners of intelligent families, to spy out the country, and on their arrival entered into active sympathy with the cause of Texas, and greatly admired Sam. Houston. Here they were, making long marches and countermarches, and getting their living principally out of the woods. We occasionally saw at this time a number of the "Telegraph," the first permanent newspaper ever published in Texas. It was first issued from San Felipe, on the tenth of October, 1835. During the Mexican invasion, it was forced to retreat to Harrisburg, and issued but one number. It was there captured by the forces of

Santa Anna. It, however, reappeared the following August at Columbia, and since that time to the present has been regularly issued. This paper was of great value to the government, and was one great instrument that led teeming thousands of emigrants to seek homes in Texas. Through this paper, which only reached us occasionally by private hands, we saw that Sam. Houston was inaugurated president of the Republic on the twenty-second day of October, 1836. Here we have his address, with these closing words, ''It now becomes my duty to make a presentation of this sword, the emblem of my past office. I have worn it with some humble pretensions in defence of my country; and should the danger of my country again call for my services, I expect to resume it, and respond to that call, if needful, with my blood and my life.'' Fired by such sentiments as these and from such a source, our company of soldiers seemed perfectly willing to fight and bear their own expenses.

Before me lies an estimate of the population of Texas made in September, 1835: —

Anglo-Americans	30,000
Mexicans	3,470
Indians	14,200

I must here be permitted to enter my protest against the correctness of this statement, so far at least as to the English-speaking population. There was at that time, as is usual in almost all new countries, a disposition to exaggeration. It was impossible then to take the census of this country, and I doubt exceedingly if the authorities would have been willing for the census at that time to have gone forth to the world under oath. During 1837 and 1838 I travelled, as will be seen, over almost all the territory of Texas then inhabited by the Americans, and I have no idea there were over fifteen thousand white people in the country. The number of Indians must have been greatly underrated in this statement.

Reaching the colony at the Falls, we found great excitement. There were six or eight families in the colony; at the fort near by the camp were some thirty or forty soldiers. So soon as things were somewhat composed, we had an appointment for preaching.

This was continued once a week, when I was at home, and circumstances would allow. We were frequently interfered with by reports that Indians were in the neighborhood. Rangers, who were kept out as spies, would very frequently come in and report smokes on the west side of the river answered by smokes on the east. Indians had set times on the frontier to move south and do mischief; this was generally on or about the full moon. Travelling, as they usually did, in separate, small detachments, with points designated at which to meet, they would frequently kindle fires and throw piles of green moss from the trees on them, and in this way, in an open country like Texas, they could easily communicate with each other. Generally the rangers would detect their advance, and we were notified, but not always.

About the first of January, 1837, I was notified by the commander of the fort that the ammunition was almost exhausted; that there were not five rounds to the man; that the government had neither money nor lead. We were of course in imminent peril. This was *our country,* and *our fight*; and although it was painful under the circumstances to leave my loved ones, exposed as they would be, my sense of duty to the land of my adoption required that I should go alone to the town of Washington, one hundred miles south on the Brazos River, in search of powder and lead, at my own charges. At this point the devil sorely tried me. The question was asked, "Now, sir, do you believe the language of the Bible, from which you preached so earnestly to the people in the old nown of Nacogdoches, just one year ago?" After a little season of meditation and prayer the language of my soul was sounded out audibly from my lips: "Yes, I believe, yet, that the wilderness of Texas will blossom as the rose, and the solitary places be made glad by the presence of the Lord;" and I started.

On my way down, travelling at one time thirty miles without seeing a human being, or even the habitation of man, my mind was active, and resolution firm, to preach whenever and wherever opportunities were offered. At Nashville I found six or eight families, and, as I must tarry there a night, I called the families together and preached.

Arrived in the little town of Washington about sunset; met a man on a crutch; inquired for the public house, and after he pointed it out he inquired if I were not a Baptist preacher. I replied that I bore the name of one in Tennessee, and would not deny it in Texas. He invited me to preach for them that night, to which I consented. This man proved afterward to be Brother N. T. Byars, who preached for so many years as a pioneer Texas missionary. Appointment was made and filled, Everybody, we were told, turned out. The room obtained was filled, and many stood outside. This was the first sermon ever preached in the town. There were then three or four Baptists there, and of course they were greatly pleased to hear the sound of a Baptist preacher's voice. I retired to rest as soon as I could, having travelled entirely alone, along crooked Indian trails, one hundred and twenty miles in two days, and preached each night.

In every store in the town I inquired next morning for powder and lead. On keg of lead was found, but no powder. As much lead as was thought to be safe was put into my saddle-bags. Several bars were bent, a string run through them, and balanced on the horn of the saddle. The way selected to return was by Independence. It looked more like dependence then than independence. Such was the appearance of all our towns. I soon reached the Yegua bottom, — not a very interesting place then, nor now, when the steam is swollen. This is one part of Texas, according to the writer's opinion, that has made very little improvement, though it be under the shadow of Baylor University. The steam was swimming, for about thirty feet, in the main channel — the whole bottom, nearly three miles wide, was a sea of water — no bridge — the horse was still able to go sixty miles a day, and must carry me by Jackson's store and to Nashville that day. He carried me over safely with the load of lead.

Jackson's store was reached late in the evening, within eight miles of Nashville. Weary and hungry and impatient, I entered the store and asked for powder. Some had just been received by a wagon from Columbia. Men were then crowded in the room, who had engaged all the powder, and paid the money in advace, previous to its arrival. After some threats made on both sides of the question — a description given of the condition at the Falls — my long trip — shivering then in my cold, wet clothing, six canisters of powder were received and paid for, and I was at home on the fourth night. Rode the same horse two hundred and forty miles inside of four days. The soldiers, on receiving the powder and lead, were in fine spirits. There was no danger of starvation, with plenty of ammunition, and hopes were entertained that the Indians could now be held in check.

There were no Baptists at the Falls, except myself, my wife and daughter. Cut off thus from all communication with churches or ministers, the situation was by no means a pleasant one. It was thirty miles to the nearest settlement on Little River. A few families were at Parker's Fort, thirty-five miles distant, near the present locality of Springfield. It was unnecessary to make appointments at these places, with any prospect of filling them. Thus cut off, I did what I could for the spiritual welfare of those by whom I was immediately surrounded.

CHAPTER V.

AN EMERGENCY. — IN 1837

OUR Indian troubles increased rapidly toward the close of 1836 and the beginning of 1837. About the first of February, 1837, a light snow covered the earth. This was rather remarkable then, and presented quite a contrast with the former season. The spies from the two forts at the Falls, and on Little River, met every day on middle ground. Reports were given, at both places, that the trail between the forts had been crossed by Indians going south-east toward Elm Creek. All the sign that had been discovered made the impression that it was only a small party on a thieving expedition. About fifteen men under Lieutenant Erath, some from each fort, met at the point where the Indians crossed, and followed the trail almost to the mouth of Elm Creek. Here suddenly a number of trails came together, showing evidently that the Indians were not few but many. They had up to this point been travelling in detachments, and it was one of these small companies whose sign had been discovered between the forts. From an elevation the rangers saw the smokes from camp-fires in the bottom. The few, however, under this brave lieutenant determined to attack the many. The horses were tied some distance out, and the fifteen men, at great peril, under cover of the night, cautiously approached the camp. Arriving at the point before day, they took shelter under the bank of the creek within thirty feet of the fires, and waited for the morning. Strange to say, the dogs of the Indians, watchful as they are, did not discover the Americans. Daylight streaked the east, and the Indians began to rise from their beds and stand about the fires. The officer had divided his men into squads of three or four, and ordered them to shoot at different fires, and to be certain that no two men were to shoot at the same Indian. Just as the last Indians were huddling around the fires, the dogs made the discovery, and directed attention to the spot where the rangers lay concealed. The command was given to fire. A number of Indians fell at every camp-fire, and there was a general stampede among the survivors. Every gun was now empty, and the Indians were thought to be about one hundred and fifty strong. Discovering that the rangers were few, they soon rallied, and the fight became a desperate one. Lieut. George Erath ordered a retreat. He kept himself in the rear as the men were getting out. Brave officers are usually in front of the charge and in the rear on a retreat. Two of the men were killed on getting out of the ravine, — Frank Childress and another whose name is not remembered. A large number of the Indians had guns. Some forty or fifty balls were afterward found in the tree at the point where the rangers first showed themselves rising from the ravine. Indians were now pressing them closely. Lieut. Erath was on one side of the ravine and a large Indian on the other; both their guns empty and both loading, glancing at each other. Erath took no time to measure his powder, poured down a handful. Both raised their guns about the same time; both fell. The Indian was dead; Erath instantly rose to his feet. A ranger cried, "George, are you hurt?" Erath, a German by birth, replied, "No, I ish not hurt; my gun knocks down before and behind." His gun had kicked him down. Thirteen American rangers escaped. Childress was not killed dead on the field, but was afterwards found, with his gun, sitting by a tree a little way off, and his body resting against the same tree lifeless. The other man was scalped, and his hands cut off. The Indians buried their dead in a pond of water and fled. They were pursued, but not overtaken. I make this statement upon the authority of the men on their return.

But for the engagement, this large body of Indians would very soon have been in the settlements below, killing, burning, and stealing; for they never came down in such large numbers in those days without desperate ends in view. It was believed to be the same body of Indians that killed poor Harvey and his family, during the month of November preceding; and, also, the same that approached the camp where I and my family lay close by Harvey's, the night previous to his murder.

Now I felt a thousand times paid for my long ride to Washington, amid so much exposure and anxiety. This work was done with the ammunition that I procured while on that trip. But I greatly lamented, in the end, the loss of the noble son of my dear sister Childress. He was, indeed, a promising boy, about eighteen years of age; was not a member of the company of rangers, but, as a citizen, volunteered his services for the trip.

It was now time to commence farming. Facilities for this were very poor. Calling to mind the events of the past few weeks, the reader will readily understand that I had the good-will of all the soldiers in the fort. We were living on the west side of the river; our little twenty-five acre rented field was on the east side. During most of the season the horses and oxen had to swim the river twice a day. We crossed in a canoe. The rangers stood guard for us, free of charge, while the hands were planting and cultivating the crop. After planting the little rented field, that was under fence, we went back further, on to Weed prairie, and planted about fifty acres with no fence. This land, with the little work we were able in primitive style to give it, made fifty bushels of corn to the acre. Portions of this prairie were cultivated for eighteen consecutive years, without fence, with nothing to disturb the corn much except the wild bear. The few cattle that were in the country went back on the prairies, during the crop season, into the higher lands.

While we were rejoicing at the prospects of the first Texas crop we had ever cultivated, the Indians were again back, and in various quarters committing depredations. Again I was informed by the captain commanding the post that ammunition was almost out, and that all their stores of every kind were very short.

The financial condition of the country was truly distressing at this period. About a half million was now due for supplies, a half million due the army and navy, and a hundred thousand dollars due on the civil list, and not a dollar to pay for anything. Near a million and a quarter of indebtedness hung over us. The families and the forts on the frontier had either to join hands, and farm and fight together, at their own charges, or fall back, leaving the people in the settlements below to be the extreme frontier on the north. Ours was the extreme northern fort at that time. I had in my possession about two thousand dollars, realized from lands sold in Mississippi, with which I expected to support my family in extreme emergencies. The officer above alluded to informed me secretly, as he did on the former occasion, of the true situation; saying that "secrecy was safety," and appealing at the same time for aid, if in my power to afford it.

Our family stores, hauled by ox-teams from Nachitoches, a distance of three hundred miles, were about exhausted, from divisions with the sick among soldiers and citizens. Although it was regarded as a savage country, yet some traces were seen of early scriptural usage. We had "ail things common." My supplies for the year would as a consequence be gone by April. The advertisements in the "Telegraph" showed that supplies might be obtained in Houston, about one hundred and sixty miles off. We determined, in this great emergency, then about the 18th of March, 1837, to attempt the trip with an ox-team, with the probability of being captured by the common enemy, and slain by the way; or, if permitted to return, with the probability of finding my little log cabin in ashes and my family having shared the fate of Harvey's, before alluded to.

Now, reader, I'll suppose you are comfortably seated on cushions, and by a warm fire, on one of those cold days in March that now sometimes are met with in Texas; that one of those splendid coaches, that is carried by the iron-clad, bellowing horse, in which you sit, is standing at Kosse, on the Central Texas track, just opposite and about sixteen miles from my log cabin in 1837, and bound for a trip to Houston and back, three hundred miles, in fifteen hours; and compare your prospects for getting supplies from Houston and mine thirty-five years ago. Two whistles, the bell rings, and you are off; my oxen, eight in number, yoked and bound with chains to a wagon that would bear five thousand pounds, my little son twelve years of age the driver, and with two horses, I am off. The road we travelled led us right along near the present Houston and Texas Central Railroad track. We camped on the ground now occupied by the city of Bryan, and within a few yards of brother W. B. Eaves' shop. Right there my horses were both stolen. Spent one day in pursuit, recovered one horse, bought another, and went on. The thief, I am sorry to say, was a Texan. Look out now, through the window, as the wheels of thunder underneath you are rolling over that magnificent bridge, that spans with its mighty arms the Navasota River, and take a look at my craft, crossing about one hundred yards above. The stream was very high, nearly level with the banks; water in the bottom almost swimming, before we reached it. We were soon through this, and on the bank of the main channel. My little son was greatly discouraged. "Pa, what are you going to do?" came out in trembling accents. "You said when we got here we would get something to eat; we have had no bread since yesterday; there is no boat, no canoe. Our skillet is lost, and we can't even get a place dry enough to bake an ash cake.' Poor boy, this trial, the hardships endured before, and those to come after on this trip and others, were preparing him for the hardships of the Mexican war in 1845, and for the successes in business that awaited him in California, where he died in 1860.

A block ten inches wide was soon hewed out from an ash-tree, and an old--fashioned "johnny-cake" baked before the fire, some meat broiled, coffee made, and father and son were merrily chatting, after a hearty meal.

I now cried at the top of my voice, to gain the attention of the man who, I was informed, lived on the hill on the south side. An answer was returned, and the man in quick time stood on the bank. "What will you have, sir?" — "I want a man to tie a rawhide lariat round that stump near where you stand, and the other end and an inch-auger brought to me, and all the assistance necessary to enable me to get my wagon and team over." Five dollars was announced as the price for which he would undertake, and the trade was closed. The end of the lariat and the inch-auger were soon brought over. At great peril, logs were cut of proper length, hauled up, and fastened securely together. On this little craft my son and I placed our camp equipage, and pulled across the river by the lariat, fastened on each bank. The man left behind then loosed his end of the rope, and, after tying the bed of the wagon fast to the wheels, tied it to the end of the tongue. The wagon was pulled over, and the team swam. By this time our contractor, one of the best swimmers I ever saw, was on the home side and ready for another contract. He then engaged to dig a canoe, by our return, for five dollars more. Here was ten dollars for crossing the Navasota twice, with great labor and peril. All this was done inside of three hours, and we were on our way through the prairie, where the city of Navasota now stands. Nothing of interest transpired till we reached Matthew Burnett's, about one mile below Cypress City. He was just from

Houston that day, and stated to me that there was no lead there; but informed me that Sam. Houston, on his retreat the year before to San Jacinto, had left a pig of lead, weighing seventy-five pounds, that was then under his house, and that I could get it on my return. Houston was reached late Saturday evening, after swimming the team over Buffalo bayou, just opposite Main street, as the city is now laid out. No bridge nor ferry-boat then. There was a little flat boat that carried over a single horse or empty wagon.

Houston, in 1837, was a city of tents; only one or two log-cabins appeared. John K. Allen's framed building was raised, covered, and partly weather-boarded. A large amount of goods in tents, a large round tent, resembling the enclosure of a circus, was used for a drinking-saloon. Plenty of "John Barley Corn" and cigars. This was the last Sunday in March, and after changing the garb of the wagoner for one similar to that worn in the city, I went out in search of a place to preach. Upon inquiry I was informed that there never had been a sermon preached in the place. It was quite a novel thing then to hear preaching, and some, to enjoy the novelty, and some no doubt with the purest motives, went to work, and very soon seats were prepared in a cool shade on that beautiful spring morning. The sermon was preached to an attentive, intelligent audience. Brother Marsh, a Baptist minister from Mississippi, then about seventy years old, came forward by request and closed the service. From near the spot where I then stood is now issued the Texas Baptist "Herald," sending streams of light over a vast territory, and around that spot has gathered a city of eighteen thousand inhabitants, destined to assume much larger proportions yet, in consequence of the fact that it is the great railroad centre of the State.

Monday morning ammunition was sought for in every store. Purchased two kegs of powder. No lead to be had. Family supplies and some additional articles for the soldiers were procured, and as rapidly as an ox-team, heavily loaded, could carry us, we made our way towards home. At Barnett's, we took in the pig of lead, a very valuable article at that time. Reaching Navasota, we found the canoe ready, according to contract, and paid five dollars for it; also five dollars additional were paid to the same man for assistance rendered in getting wagon, team, and cargo across the river. Here, you will remember, fifteen dollars have been paid for ferriage, going and coming, over one stream. Texans would do well to bear this in mind, among the charges they so often prefer against the company that drives the iron horse, with his huge trains of carriages, and in such quick time brings their supplies. On reaching the present locality of Bryan, my other stolen horse was found, and twenty-five dollars' reward paid for that. When we reached the eastern bank of the Brazos, opposite our log cabin and the soldiers' fort on the west, and announced across the river that we had powder, lead and commissary stores, hats were waved, and as loud a shout was raised as would have been during the late war on the arrival of a seventy-four gun ship in some great emergency.

CHAPTER VI.

LIGHT AHEAD. — 1837.

LIFE then as now was a mixture of joys and sorrows. The little colonies here and there were frequently greatly elated, catching eagerly at every little ray of light that made their prospects even tolerable. Oftener, however, gloom hung over the camp. From north, east and west, rumors reached us of Indian outrages, that made our blood chill sometimes with fear, and then by turns boil with feelings of furious revenge. While rejoicing with the family and soldiers over our safe return from Houston, and the bountiful supplies for a few months at least, news reached us that the little daughter of Harvey, whose sad fate has been recorded, was alive in Mexico, and a negro girl whose life was spared by the Indians in the midst of the same massacre. Great anxiety was felt by us all for the rescue of the child.

Her uncle, James Talbert, was then living in Alabama. After long search and a large expenditure of money, this brother in Christ found the child. She had been sold by the Indians, and was now greatly attached to her Mexican mother. Her arm had been broken during the killing of her parents. She was carried by the uncle to Alabama, and by him was afterwards brought to Texas. They settled near where her parents were killed. She has since married, and when recently heard from was living. I have often since been at her house and used the family Bible at worship owned by her father, and which yet has upon its pages the blood of her parents, spilled by the hands of the Indians on that fearful night.

Our opportunities for preaching were very limited. Our crop was cultivated in 1837 under a guard of soldiers. In a short time I ventured down to Nashville, forty-five miles down the river, and all the people in the settlement that could, turned out to preaching in a little log cabin, with dirt floor. Just about the time we closed the services on Sunday, the Indians dashed upon us and killed two men, in sight of the congregation. Preacher and people carried carnal weapons with them to the house of God in those days, and did not for a moment suppose they were violating the Scriptures. We instantly changed the services into war with the Indians. Every man was immediately mounted and off with gun in hand on Sunday evening, in full pursuit of the Indians. They were not overtaken, but escaped up Little River.

Now please remember my situation on the return from this pursuit. The relatives of the dead are in tears, and at their request I must stay and perform the funeral services. The Indians had gone towards my home and loved ones, apparently intending to bear a little to the left; but we never could tell where an Indian would turn up next. Duties to the bereaved being performed, and the dead buried out of their sight, I resolved to go home, if God would spare my life, under cover of the night. Forty-five miles to ride alone in the night, with a knowledge of the fact before me that the Indians were above and between me and my home, and that I was liable to be attacked at any moment. This state of things presented a strong temptation. On my way in the night, and at the crossing of Litte River, the tempter came, and his speech was as follows: "Where is your faith now? You had better accede to the propositions of your very liberal brethren in Tennessee and Mississippi, even if you die, than undergo such hardships by day and by night. The Indians will certainly get you yet, either on this trip or some other." By this plausible story I felt for the time being influenced, and there was a little wavering. God, I thought, certainly had made it plainly my duty to live and labor in Texas, and, with prayer for divine aid, my mind dwelt upon the deliverances of old, and was greatly strengthened. God sent an angel to provide food for Elijah under the juniper-tree, when he had despaired and was willing to die, and sent fire to consume the offering in the presence of the prophets of Baal. He also put into the heart of Rahab the harlot to conceal the spies while examining the city of Jericho and its fortifications, and finally caused its walls to tumble down at the sound of the rams' horns, leading Israel safely into the promised land of Canaan. Amidst these meditations I could but say, "My heart is fixed." God gave me an inward token that I should be concealed from the Indian's watchful eye, and that he would recognize my offerings in years to come. The wilderness would yet blossom as the rose. In safety about daylight I reached my home and loved ones, and found all well.

The supplies procured in Houston, on my former trip, were about exhausted in May, and it was necessary for me to return with my wagon to the city. There was neither gold nor silver in circulation then. Texas did not have a supply of "red-backs," as was afterwards the case. Our currency consisted of paper from the banks of the older States. On my arrival in Houston I found business stagnant, in consequence of the failures on the part of the banks referred to. My money on hand was all declared worthless, or nearly so, and three-fourths of the people in the State were in my condition. The first of June I was in the city of New Orleans, and walked the streets of the Crescent City for fourteen consecutive days, trying to make negotiations. It was then, with many of us, a bread and meat question, I was introduced to some gentlemen in debt to the banks, and succeeded in exchanging my bills for goods, at about sixty cents on the dollar. I had not come to Texas to sell goods; but, to save the little means left, I found myself unavoidably drifting into merchandise. As rapidly as possible I made my way to Houston with the goods that had been secured.

On my arrival letters were received from home, giving information that the Indians and Mexicans, about the middle of June, had overpowered the fort at the Fallas, and had killed Coryell, one of the rangers. My family had been run from home, and were at Nashville, forty-five miles down the river. My wagons were from home, and nearly everything, in the way of provisions and household furniture, had fallen into the enemy's hands. Truly it seemed that all my misfortunes were coming upon me at once. God be praised, however, my wife and children lived. My goods were stored in Houston, except two wagon-loads that were carried to Washington as soon as possible. My family was brought to Washington, and we were soon in business, trying to recuperate after so much loss. The best crop I ever made was all lost, our household furniture and farming tools all captured, and about a thousand dollars lost in the failure of the banks. The goods were all sold out as rapidly as they could be brought from Houston.

During this time a prayer-meeting was organized at Washington, and we found present the following Baptists to take part: H. R. Cartmell, a deacon from the Baptist church at Nashville, Tennessee; A. Buffington and wife, and N. T. Byars, also Richard Ellis and brother Jenkins. These, with my own family at Washington, brother Jesse Webb, living near the present locality of Calvert, sister Childress, at Nashville, and sister Hall, near where Chappell Hill now stands, were all the Baptists I then knew. There was an organization of some ten or twelve members east of the Trinity, near Nacogdoches, under the pastoral care of Daniel Parker, of "two-seed" notoriety. There was also an organization calling themselves "Primitive Baptists," on the Colorado River, twelve miles below Bastrop, and under the pastoral care of Abner Smith. Elder Isaac Crouch was a member of the body. This brother, sound in the faith, and of more than ordinary ability as a preacher, afterwards moved to Nashville, and in the spring of 1836, if I remember correctly, was killed by the Indians, about a mile and a half from the present locality of the Little River Baptist church, in Milam County. I was well acquainted with brother Crouch, when he was the Moderator of the Big Hatchie Association, in Tennessee, and was with him five years before at the association, when held with Mount Pleasant church, twelve miles south of Boliver, Hardiman County. This association, at that time, was largely anti-missionary. In 1825, Elder Joseph Bays preached at the house of Moses Shipman, west of the Brazos River. He afterwards moved into Eastern Texas, and labored in the vicinity of San Augustine, where he met with violent opposition on the part of the authorities, and was for some time greatly hindered in his work. In 1829, Elder Thomas Hanks preached in the house of Moses Shipman, before mentioned, under whose ministry the wife of James Allcorn, a deacon, made a profession of religion. This was the first conversion we have any knowledge of.

Brother Hanks was from Tennessee, where the writer knew him well, having labored much with him in my early ministry. During the same year, the Baptists organized the first Sunday school in Texas, with brother T. J. Pilgrim, now living at Gonzales, as superintendent. Baptists have often been charged with indifference about the spiritual welfare of children and youth, because they refuse to have their children sprinkled. Their interests in the Sunday-school cause in Texas, and elsewhere, is a sufficient defence against this charge. There was, sometime previous to the declaration of Texas independence, a pious sister, living near Gonzales, named Echols, who was a devoted Baptist, and loved her Bible dearly. The Mexican government was under Catholic rule, and, of course, the Bible was prohibited from the people; severe penalties were annexed. This sister, on one occasion, seeing the Mexican justice approaching, was tempted to conceal her Bible, that was then open by her side. Committing her way to God, she reconsidered the matter, and determined to risk the consequences. Witnessing her devotion to the book of God, his heart failed him, and he allowed her to keep it and enjoy it. There was, also, sister Mercer, the wife of Eli Mercer, living east of the Colorado, fifteen miles above Wharton. Dr. Marsh, a Baptist minister, advanced in age, alluded to as closing the services after my first sermon in Houston, settled on the San Jacinto, and afterwards returned to Mississippi and died. This, according to my recollection, embraces all the Baptists in Texas up to 1837.

In May, 1837, on my second trip to Houston, a "Committee of Vigilance" was organized, consisting of the following ministers; Dr. Marsh, Baptist, Dr. Smith, Protestant Methodist, Rev. W. W. Hall, Presbyterian, Rev. L. L. Allen, Episcopal Methodist, Rev. H. Mathews, Episcopal Methodist, and Z. N. Morrell, Baptist. The object of this committee was to prevent impostors from securing the confidence of the people, in the exercise of ministerial functions. A notice of this organization, with an address to all ministers coming to Texas, was published in the "Telegraph." After this no minister was recognized by any of us, until he exhibited unmistakable credentials of authority from his denomination. The effect was very beneficial to the cause of religion in the republic.

The circumstances that gave rise to this organization were as follows: An imposter, a short time before, came to the town of Washington, in my absence, and preached under Baptist colors. He represented himself as sadly in need of pecuniary assistance. Some person kindly circulated a subscription, which procured the amount of funds called for. He was soon seen at the grocery, and then on the race-track, as one of the standard-bearers of that immense number of sportsmen that gathered about the town of Washington. I would be glad to know that every Baptist pastor in Texas at the present time considered himself as a member of a Vigilance Committee. An impostor, if he succeed, is a great curse in any community, and should be persistently ruled out. Impositions have been practised in Texas since 1837, and may, in the future, without great care. No pastor should allow a man in his pulpit until he gives clear evidence of his right. He who takes offence because his papers are demanded, is, to say the least, a suspicious character. Ministers are required to be "of good report among them that are without," and where there is the smallest ground for suspicion, whether on account of being a stranger or otherwise, an opportunity will be given by such investigation for a good man to set himself in order before the people. Good men, as preachers, have often been seriously injured because this course was not rigidly pursued.

The finances of the government were now in a very deplorable condition. General Houston, desirous of incurring as little other indebtedness as possible, was anxious to discharge a large part of the Texan army. Having nothing to pay them in that event, he resolved to follow the precedent laid at the close of the revolution of 1776, and furloughed a large number; it was supposed about two-thirds. Those who had families in Texas quietly repaired to their homes, and entered upon the common avocations of life. There was another class of these soldiers, with whom we sympathized greatly, and yet they proved a terror to good society. They were principally young men of the very first families of the United States, mostly from Tennessee, on account of the popularity of Sam. Houston, once the honored Governor of the old State, then President of the "Lone Star Republic." They were young men deeply imbued with that spirit of patriotism that fired the hearts of the fathers of 1776. They had now served in the Texas army for several months; there was not a dollar with which to pay them for service rendered; their clothing was worn to tatters; they had not been accustomed to labor, and were now a great way from home, and all their available means consisted of land certificates and bounty claims on the infant republic. Large bands of them were among us, fit subjects for every species of dissipation. The great dreamer, John Bunyan, very truly said that "an idle brain was the devil's workshop." The sequel will show whether this be true or not.

Our weekly prayer-meeting was regularly held in the town of Washington, in a small house, the best we could secure. These young men referred to, regularly attended; behavior was good; were very polite, and sang elegantly all the parts of music. They had been trained to this in other States, under pious influences. A stranger present would have supposed that a whole church, well-organized and drilled in some of the old States, had moved in a body and settled at Washington. Cartmell, Buffington, Byers, Ellis, and Morrell, one after another, led in prayer, and the singing between prayers was of the very first order, in point of time and melody. The writer would give out an appointment for preaching every Sunday, when at home, and after singing "Old Hundred" the congregation would retire. After the benediction the young men would hasten away. By the time we would pass on our way home, the grocery and billiard-saloon would be lighted up, and a large crowd — God have mercy on them! — would be assembled for the night. Here was an important move of the prince of darkness, — his image and sign hanging over the door. There was King John Barleycorn within, double-refined, with all his machinery propelled by the engine of hell, fed with the fire of damnation, drawn directly from the "bottomless pit" of eternal perdition. It did not require the foresight of a prophet to understand the results of this procedure, on the part of the enemy, if continued long. Our prayers went up to God, "O Lord, hear the prayers of thy servants, and the prayers of mothers, in distant lands, for these wayward sons!"

We determined, let come what might, to organize a church. The day was appointed, and eight Baptists assembled to keep house for God. Brother H. R. Cartmell was recognized as deacon, and Z. N. Morrell chosen as pastor. Thus sprung into existence the first church, according to my information, that was ever organized in Texas on strictly gospel principles, having the ordinances and officers of ancient order, and with no anti-missionary element in its body. Shall we be permitted by the enemy to remain together, and enjoy church privileges? We had long desired such an estate. Shall the present organization stand, as the nucleus around which others will grow up? This we fondly hoped. Or shall the feeble light in the wilderness be blown out, and require resuscitation? Such questions, in those troublous times, revolved in our minds. We must wait and see.

A committee consisting of J. R. Jenkins, A. Buffington, and H. R. Cartmell, was appointed to correspond with the mission boards, north and east,and request that Texas be taken into consideration as a missionary field. An appointment was soon tendered to me from the American Board. Good reasons, at the time, were between me and the acceptance of this proposition. The correspondence was continued, and the result was, that Elder Wm. Tryon was sent to Washington County, and Elder James Huckins to Galveston. These brethren did not come for several years after this correspondence was commenced. Of them we will speak again at the proper time and place.

Measures were at once taken, by the infant church, to build a house of worship. The money was subscribed, the material was secured, the building and the work went rapidly forward towards its completion.

In the fall of 1837, some families were passing through the town of Washington, on their way to Gonzales, who had left during the retreat the previous year. Among them was a very intelligent couple, who requested, at my hands, the performance of the marriage ceremony. This was the first time I solemnized the holy rite in Texas.

I was much interested, as they related the existence of their long engagement, and their unwillingness to be married by a Catholic priest, according to Mexican law. They had patiently waited till the congress of Texas passed laws, and marriage license could be issued by authority of the republic.

Previous to the independence of Texas, marriage was illegal, performed by any save a priest. Catholic priests were very offensive to Texans, and for the performance of the ceremony they exacted twenty-five dollars. Many refused to submit, and, in some such cases, the parties simply signed a bond in the presence of witnesses, and became husband and wife.

The congress very soon passed a law allowing these parties to take out a license in due form, and be married by a proper officer. When the license and bond were returned, with the certificate of the officer performing the ceremony, the marriage was legal.

I was called on frequently afterwards to officate in such cases, and, in a few instances, a group of little children were witnesses for their parents. In one instance, immediately after preaching, I performed a ceremony in the presence of the congregation, the parties each holding a child in their arms.

CHAPTER VII.

FIRST FRUITS IN THE MIDST OF TRIALS. — 1838.

GOD has determined that his saints shall come up to him "out of great tribulation, having washed their robes and made them white in the blood of the Lamb." In keeping with his general providence it seemed to him necessary that the early Christians of Texas should be sorely tried. Such was the experience of the brethren composing the little church.

The furloughed soldiers, without any employment, were daily becoming more dispirited and desperate. It required all the ingenuity of General Houston to retain them in the country, and prevent discouragements on every hand. A fine opportunity now offered itself for the display of that generalship that has rendered his name immortal. Threats were about this time made by Santa Anna, that the Texan rebels should be driven out, and that the flag of Mexico should be planted on the Sabine. He threatened, also, retaliation upon the authorities that had encouraged the immigration of the young men above referred to. These threats reached President Houston just at the time to answer his purposes best. He deliberated a day or two, and dipping his pen into an inkstand richly tinctured with sarcasm, that he always kept on hand, he wrote Santa Anna a letter through the public press. The substance of the letter was, that Santa Anna, his former prisoner, would accommodate him very much if he would concentrate the flower of the Mexican army in large numbers, mounted on Mexico's best horses; if he would drain the Mexican treasury in procuring a good outfit; and if he would draw largely upon the priests for money to meet the expenses of the campaign; promising him that if he would cross the Rio Grande with his army, the Texans would wipe it out of existence, and thus secure an outfit with which he would march the Texan army through Mexico, and plant the colors of the "Lone Star Republic" on the isthmus of Darien.

General Houston cared more for the effect this letter would have on the furloughed soldiers than upon the man to whom it was addressed. Upon these men he must depend to meet and repulse the Mexican raiding parties that he expected would harass the country in connection with the Indians. This communication was a fine feast for our Washington gentlemen, and they stood ready for the emergency, at their own expense. While their admiration for Texas and Sam. Houston was being fired, their morals were being daily corrupted. The plainest evidences of the fact now became manifest to all.

Our meetings on Sunday were very regularly kept up, and the prayer-meetings continued.

About midway between my residence and the little house of worship, and about sixty yards from each place, in the principal grocery in town, an opposition prayer-meeting was organized. At first they did not interfere with our meeting. All the crowd would attend ours, and immediately after it was closed they would gather in the grocery, and the leader of the band would open in due form. Our services were imitated to the best of their ability. Our names, who led at public prayer, were called at the grocery with a loud, clear voice, and parties there responded with prayers and exhortations. Under this trial the tempter sorely distressed me; sometimes with a spirit like that of Peter when he smote off the servant's ear; sometimes with misgivings; and sometimes with the pains of deep mortification and grief. Under all this, kept up regularly during successive weeks, I resorted much to prayer around my family altar, and in secret. There was no law then that we could use to break down this great evil, that was so fearfully contagious in its character. The bread and wine, emblems of a Saviour's love, were frequently administered by these mockers of God and religion, before the public gaze. The only consolation I could derive was that of which Paul spoke in his letter to the Philippians. Christ was preached, whether sincerely or otherwise, and although they supposed "to add affliction to my bonds," I prayed that they might at last, in the midst of their mockery, be able to look through the mirror of the gospel, become alarmed, and repent.

Those are fearful days for any people when an army is disbanded among them, whether that army has been successful or unsuccessful. Many citizens of this great nation can testify to the truth of this assertion, in more instances than one, from personal experience and observation. In the fall and winter of 1837, quite a change for the better came over the country. A large number of immigrants came into the republic; good crops had been made; our financial condition was somewhat improved, and the laws passed by the congress were being put into execution to some extent. The year 1838 opened with flattering prospects, compared with the past. The little town of Washington was incorporated, and some imperfect police regulations were established, requiring a small taxation of the citizens. All this, however, had but little effect at first to restrain idle soldiers that still hung around us. Gambling and drunkenness were carried to a fearful extent, in spite of all the restraints of our imperfect social organizations and civil regulations.

Elder Robert Alexander, the first missionary from the Methodist Episcopal church south, had come to Washington the previous year, and had preached on several occasions. Dr. Smith, a Protestant Methodist, had fallen in among us.

Two Cumberland Presbyterian preachers arrived, — Roark and Andrew McGowan. The latter I had met in the spring of 1836, on his way from the San Jacinto battle-field. From his messmates I had learned that he went through the campaign, and preserved his Christian integrity. In my opinion, then, and ever since, this is quite a recommenation to any man.

These preachers were now present, intending to hold a protracted meeting. This was the first meeting of days ever held in the town, and it was rather more than the fiends and mockers could willingly submit to. The house in which they proposed to hold the meeting was a vacated billiard-room on Main Street, with a long gallery in front. On the second night of the meeting there was a general attendance of the citizens, loafers and gamblers of the place. We soon discovered that the disturbers of our peace on former occasions were present, with the intention of interfering with the worship of the congregation, without the fear of God or man before their eyes. A man was stationed outside of the house, just behind where the preacher stood, with a hen in his arms. While the preacher was lining out his hymn he would hold the chicken by the neck. When the congregation would sing he would make it squall. A large copper-colored negro man was stationed on the gallery in front, with some twenty or more of these lewd fellows around him, partly intoxicated. When the congregation sang and the hen squalled, the negro, acting under orders, would put his head in at the window and shout at the top of his voice, "Glory to God!" The response from outside was given, "Amen and amen!" I was sitting near by the window from whence the disturbance came; my wife and daughter were near by me. I arose and stood by the window with the walking-cane in my hand that I had brought from Tennessee, made of hickory, with a buck-horn head. My bosom heaved with holy indignation, and as the negro put his head into the window the second time, as the congregation sang and the hen squalled, I struck him just above the left eye, making a scar that he carried to his grave. This band had always treated me with courtesy, yet it was clear to my mind that they intended to drive these preachers from the town, and I felt confident my time would come next. After the stroke with my cane, they were peremptorily ordered away, with the statement that there were more dangerous weapons than the stick behind. It had been customary with us, since the Indians killed two of our men during religious service at Nashville the year before, to take our weapons with us to church, as well as to other places. Some usually stood guard while others worshipped. There was no further disturbance of consequence until the services were over. The sermon was preached by Mr. Roark; Mr. Alexander closed.

Before the congregation was dismissed, I claimed the right to make a short but plain speech. In this speech, I stated that I had often tendered my thanks to the people of the town for their politeness and good behavior in the house of God, — regretted that the thanks tendered on other occasions were not due on this. Before me are sons from the battle-field of San Jacinto, coming from the various parts of the United States. For what did you traverse the prairies of the west, under the command of the gallant Houston? And for what did you charge the enemy's cannon and burn the bridges behind him, unless it was for civil and religious liberty? Santa Anna has been captured, and priestcraft driven from the land; and yet, in less than two years, you have commenced to pull down what you have built up by so much toil and sacrifice. We are determined, as ministers of the gospel, that we will not be run out of Texas, nor out of this town. For one I can say, let Texas rise or fall, live or die, her fate shall be mine; and I believe God will yet overrule all this to his glory. I have looked for something in the Scriptures to justify my hasty conduct on this occasion. The Saviour, driving the thieves from the temple, is the nearest I can find. In this case the house of God was made the house of mockery.

After the congregation was dismissed, fears were entertained by my friends for my personal safety. The band of mockers hung round the door to the last. Col. Matthew Caldwell, who, at the head of his command, distinguished himself on so many hard-fought battle-grounds against both Mexicans and Indians, was present with his family on this occasion. He stopped at the door before passing out, and addressed the miserable crew: "Gentlemen, I have a wife and daughters here, as well as Mr. Morrell, and this state of things shall be broken up. If there is any fighting to be done, you can put me down on the side of civilization and religious liberty." No violence was attempted upon any of us; but quite a crowd of these men followed close upon the heels of the preachers, as they retired, and barked at them like dogs.

A feeling of righteous indignation was felt in the bosom of every worthy citizen of the place, and the community was called together the next morning, at the insistance of Col. Caldwell. At this meeting, and in the presence of these ministers, that had labored for us the previous evening, resolutions were offered and passed, condemning in severe terms the manifestations and interruptions of this wicked crew on former occasions, and strongly in favor of morality and social order. I have lived in Texas thirty-four years since then, and have witnessed no more such demonstrations.

The meeting in the old billiard room continued for several days, without any molestation whatever. About the third or fourth day, one of the preachers informed me that they would be compelled to close the meeting and go home, as they only had means enough to pay their bills at the hotel, and to bear their expenses home. This was all provided for in Texas style; the horses were sent to grass, and after the bills at the hotel were paid the preachers were all provided for at private houses. Two considerations influenced us to urge the continuance of the meeting. One was in effect of the gospel upon the community at large; and the other was, that the former disturbers of our peace might understand that preachers, under the law of God and the land, had a right to come and preach in the town of Washington, as long as it suited their inclination. At the close of the meeting a contribution was made by the citizens, and an invitation extended to return as often as they could.

These gamblers have some redeeming traits of character, which I will now illustrate by a single incident, which occurred shortly after the events just recorded. These troublesome characters were all sportsmen and generally, wherever I met them then and since, would give liberally to any charitable object.

A family came out to Texas in 1838, from one of the Northern States. The husband and father died with the yellow fever on the way from Houston to Washington. The widow, with four or five children, arrived in town, and notified us that the money intended to pay the freight on their household goods was expended in providing for their sick, and in burying the deceased husband. Here was a case demanding the sympathy and active aid of the servants of Christ. There were only a few Christian people in the place at the time, and we needed more money than they were well able to give. Looking over the list, I found four Baptists, one Methodist, and one Presbyterian. Others had been there, but were not present that day. Two subscriptions were made out; one to be circulated among the moral and religious men of the community, and one for that numerous band that were on some occasions such a terror to us. Our issue made with them touching the public worship seemed to have created no alienation of feeling on their part. They met us and passed us as politely and kindly as ever. The king of the band was Captain James Cook, who had proved himself a man of great daring at the battle of San Jacinto. In a skirmish, previous to the main battle, Captain Cook's horse ran off with him, dashed through the Mexican lines, and was checked up behind the enemy's breastworks. He saw no chance to escape the dreaded Mexican prison but to charge back through the same lines. With all the dignity of a commanding general, he straightened himself in his stirrups, and ordered Celum, in hearing of both Mexicans and Texans, to "charge!" A volley was fired, taking no effect upon the horse or his rider; the line gave way, and the captain was soon with his command. The subscription was given to him, with the statement that his brethren were more numerous than ours, and that the widow was in great distress. Sixteen hundred dollars had been stolen from me the week before, and this I ventured to state to him was in the possession of some of his men. Giving this matter my personal attention, I only raised twenty dollars. Captain Cook raised eighty dollars in a very short time, and left it at the store, with his compliments to the poor widow and orphans, stating that if more was needed more could be had from the same source. He said further, addressing himself to me, "That fellow that stole your sixteen hundred dollars came right over and lost it all among us. We were not aware of the fact at the time: but had it not been for that we would all have been out of money by this time. We of course can afford to be liberal under the circumstances."

The Indians still continued their savage work. A lady, by the name of Taylor, living near the present locality of Anderson, was waylaid and killed by the Indians, while on her way to the place where her husband had been killed by them, only a short time before. Her friends protested against her expedition, but so anxious was she to look after the remains of her husband that she risked and lost her life in the effort. She was killed near the spot now occupied by the Oakland Baptist house of worship, in Grimes County. This created quite a panic. Much was said, but from some cause but little was accomplished in the pursuit. The settlements were small and much scattered; and it was the policy of the Indians, when pursued after depredations were committed, to divide out singly, and into such small companies as to make it quite difficult to follow their trails.

The little band of Baptists at Washington still kept up the regular meetings. The pastor, the writer, was by this time exceedingly tired of his merchandise. He did not come to Texas for trade. The failures of the banks, before stated, had apparently made it necessary. Upon casting up my accounts I found myself in possession of as much money, after all my losses by theft and otherwise, as I had originally, and now determined to give the business up. Selling goods don't suit a preacher, for a great many reasons. Dealing with so many people, and charged oftentimes with dishonesty, whether guilty or not, it is very difficult to preserve the qualification requiring that he be "of good report among them that are without." Besides all this, the great tax upon the time and mind, to keep a business in good condition, interferes very materially with the preacher, who is required to give himself "wholly" to his work. Peter J. Willis, now one of the first merchants of Galveston, was at that time my clerk, and to him I sold my remnant of goods, on a credit, at cost and ten per cent. Taking the business in hand, he paid me promptly for the goods, and by dint of good management and hard labor has reached his present position. Thus my days as a merchant closed, never to return.

No language can describe the great anxiety I felt to meet with a mission-loving Baptist preacher, holding, as I believed our little band did, the ordinances of the gospel as they were delivered. None came, although I had been in Texas nearly two years. As evidence of the fact that ministers need ministerial influence and aid in their work, Christ sent the early preachers out "two and two."

The summer season was passing away; Doctor Manly, a Methodist minister, was daily visiting my house, nursing and giving a physician's attention to my family, several of them sick, and my youngest daughter very low with the common fever of the country. A pressing invitation was received from Rev. Robert Alexander, to go down the Brazos River about twenty-five miles, and meet him at a camp-meeting. We differed widely on the questions of baptism, church order, and communion; but as my business was all given up, and no Baptist ministers in the country to confer with, I considered this a good opportunity for me to go and preach Jesus. Dr. Manly agreed to take charge of my family.

Everything was arranged at the place in the old-fashioned camp-meeting style. Some of the campers were thirty miles from home. Although I left home after twelve o'clock on Saturday, I reached the place in time and preached at night. As it was necessary for me to return to my family on Sunday evening, I was requested to preach Sunday at eleven o'clock. My text for the occasion was from Rom. vi. 23: "The wages of sin is death: but the gift of God is eternal life, through Jesus Christ our Lord." I must confess that I felt somewhat cast down. During the morning I had reviewed the past. Fourteen fruitful years had been gone over in Tennessee, amidst the scenes of an active and successful ministry, and now nearly two years have passed in Texas, and if a single soul had been converted under my ministry I did not know it. I consoled myself somewhat with the thought, that amidst this darkness my motives were right, and that duty had been done with the lights then before me. My line of argument drawn from the text was, that while death was the wages of sin, God was a just pay-master, and would always pay wages where and when they were due. A fervent appeal was made, showing that it was a fearful thing to fall into the hands of such a God as our God, without the righteousness of his Son. A man by the name of Jackson was sitting right in front of me, who attracted my attention by his fine appearance and marked interest in the discourse. He had come some distance to the meeting, with his saddlebags loaded with whiskey, and confidently expecting to have a mery time around the camp with his friends. Greatly alarmed at the judgments of God declared respecting the sinner at the great day of account, he sat trembling for a moment, and fell on his face in the aisle. The friends who gathered round him were admonished of the fact, that the mind for the time being had overpowered the body, and that God would take care of his soul. Very soon he professed a hope in Christ, and lived a consistent Christian to the close of his life, some years afterwards. At the conclusion of the sermon old sister Hall, a Baptist from Missouri, who had not heard a sermon for six years, and several others, praised God aloud, as we are informed they did in ancient times. To God be all the glory. A great burden was lifted from my mind, and I determined more resolutely than ever to "spend and be spent" among the people of Texas. This was the first testimony, clear and decided, that God had given in the wilderness of the west, of salvation wrought under my ministry. My sick family demanded my presence at home, and Sunday night found me watching with the loved ones there. God saw fit in his providence to spare life and restore health.

CHAPTER VIII.

THE WEST. — 1838.

HAVING glanced at the progress of the cause of Christ in 1838, in the preceding chapter, with the conflicts and trials of his scattered sheep, we purpose in this review of some other facts relative to the state of the country. Warlike demonstrations, as we have before remarked, were pointed at us, both from the north by the Indians, and from the west by the Mexicans. Under all this pressure, a large tide of immigration was constantly flowing into the State.

About the first of February, 1838, the Texas land office was opened; and be it remembered, that it was done over the veto of that general and statesman, Sam. Houston, then the president of the Republic, who by some means was wonderfully gifted with an ability to guess well, to say the least of it, as to future results. He had, with a wise forecast, seen the "Lone Star Republic" rising like a star in the distance, at the time he left Tennessee, and now he declares in his protest, that if the land office was opened, and certificates were located before the country was sectionized, it must inevitably result in an untold number of lawsuits, furnishing no good to any, save the swarm of petty lawyers and swindlers that were then in the country, and others who by the act would be encouraged to come. The truth of this prediction has been realized in almost every court in Texas for the past quarter of a century. The prediction, however, was no barrier in the way of the coming population. Certificates were bought up and located rapidly, every man supposing that he could locate and sell lands enough to realize a fortune before litigations would spring up. Some succeeded and many failed. Among those who failed were many of a class, who, on their arrival, supposed that Texans were a back-woods, illiterate people, and that they, the new comers, were members of the "King Know-All" family. Such men might be met with every day in fine spirits, and talking wisely about annexation and the future glory of the Texas star. It was the pride of the furloughed soldiers to swindle them out of all they had, and then laugh at them as "green from the States."

Land speculators swarmed like locusts. Large amounts of money were invested in certificates, and reports of the great value of lands in the valleys of the west led these men in large bodies to travel towards sunset. These speculators with guns and camp equipage were considered by the Indians men of war, and every man's path was full of danger.

Be it remembered that it was now the first of March, 1838, previous to both the protracted meetings referred to, and just after I had closed my merchandise. My lungs were quite feeble; the salt atmosphere of our south-western coast was considered as beneficial to one in my condition. Outside of the little town of Washington, I seldom had opportunities to preach; and with a slight attack of Texas land fever, so common in the country, I, too, invested some money in certificates and started west.

We have heretofore given an imperfect sketch, morally and physically, of the country east of the Brazos, and along its rich valley from its mouth as high up as Marlin, in Falls County. By consent, the territory west of this river is denominated Western Texas, and this occupies by far the larger part of the State. Look on your map, and see what a vast territory stretches away to the south-west from the town of Washington on the Brazos. In 1838, there were only a few very small settlements scattered along the Colorado Valley, fifty miles west of the Brazos. For about eighty miles west of the Colorado, the settlements were smaller and more scattered, as far as Goliad, and beyond this, in a south-westerly direction, there was not a community of civilization east of the Rio Grande. Even Mexicans feared to go through these vast plains, given up, as they were, to various roving tribes of Indians. A few of us in Texas at the time preferred to believed that these wild solitudes of nature would be eventually reclaimed, and, acting under this belief, made the necessary preparations to explore the south--west.

As far as the town of Columbus, on the Colorado, I wended my way, a distance of seventy-five miles, entirely alone. About the tenth of March, three gentlemen, according to promise, met me at Columbus, armed and equipped in ancient Texas style. Thus far there was only occasionally a house; the greater part of the country was prairie of the first quality, with a sufficiency of timber for ordinary purposes. The grass by this time was very fine, and as we travelled over the high and gently undulating prairies that lay between us and the Guadalupe River, we praised God that there was even a distant prospect that this beautiful land would one day be dotted with farms, school-houses, and edifices dedicated to God. The largest herds of deer we had ever seen appeared in every direction, as many as one hundred were in plain view at a time. We camped near the town of Victoria.

Twenty-five miles brought us to Goliad. Here stood the breastworks of 1836. We did not look on these ruins with the feelings that are experienced by the traveller in 1872. The capture and murder of Fannin and his noble band had occurred just two years before, and the relation that Texas bore to Mexico in 1838 was very different from what it is now. It was hard for the Christian spirit to maintain its sway in our bosoms. In the face of the evidences of this great outrage upon humanity we were strongly inclined to cry for vengeance. We then thought and still think that it was an evidence of great folly that Fannin should have attempted defence on such ground, behind such breastworks, and with such a small band, against such overwhelming numbers on the side of the Mexicans. Here, after a hard struggle, three hundred and fifty-seven brave Texans, two years before, surrendered, with a written agreement that they were to be treated as prisoners of war, according to the usages of civilized nations. On Sunday morning, March 27, 1836, they were led out and shot, — only twenty-seven making their escape, — leaving three hundred and thirty who were butchered in cold blood.

We found at Goliad two or three Mexican families, and about as many Irish. I secured the use of the old Catholic house of worship that stood close by the breastworks, and preached to less than a dozen persons. This was the last settlement in the direction we were travelling, and the only opportunity I had to preach on the whole trip. This was the first gospel sermon they had ever heard, and in all probability it was the last.

We were informed that these were the last white faces we would likely see before reaching San Patricio, and that there was no road across the country. The distance to San Patricio was sixty miles. The map of Texas was spread out, the compass was laid on it, and the direction ascertained. The compass, it was agreed, should be our path. The land along our way was very rich; timber and water exceedingly scarce.

On the second day after leaving Goliad, we saw, for the first time, a herd of antelopes. There was no time for conference as to what we should do. I had no intention whatever of running my horse after such game as this; but when my comrades started in the chase, and I saw the elegance with which the antelope moves, the excitement was contagious, and dropping my baggage on the prairie, I passed my associates full four hundred yards in a mile and a half, and succeeded in cutting off a young one from the main herd. Seeing that it was cut off, it fell down and cried like a lamb. Very soon I held the beautiful little animal in my hands, and would freely at that moment have paid one hundred dollars for the privilege of handing it over to my little daughter at home. We examined it well, and after our curiosity was gratified, as we could do nothing with it, it was released, with a crop and under-bit in the right and a swallow-fork in the left. Our interest in the antelope stock, we suppose, is still in the range.

It was sixty miles from Goliad to San Patricio; yet, desirous of seeing as much of the country as possible, we had travelled towards the junction of the Rio Frio and Nueces Rivers. These flow together, both lovely streams, and flow to the gulf under the name of Nueces. The junction is about fifty miles from Goliad, and San Patricio is on this river about fifty miles below. At this junction we camped, and had the greatest quantity of wild meat at our command. Our route was along the valley of this beautiful stream. Occasionally we saw signs of the Red Man, which kept us on the alert. Indians seldom give warning, striking when and where you least expect it.

We observed, while in this valley, a striking contrast between Indians and rattlesnakes. A large pile of trash was observed, full eighteen inches high, curiously placed together. We at once commenced an investigation, and before many sticks were stirred, two grand, bold, spotted rattlesnakes made their appearance. Without the least attempt to take us by surprise, the war-drums on the ends of their tails struck a few sharp notes, the weapons of war were freely exposed, and with heads erect above a splendid coil, they hissed a challenge for a fight. The challenge was accepted, but not until we had secured poles that we supposed would measure about double the length of either snake. After a short engagement, none of our men being killed or wounded, the snakes lay dead at our feet, measuring each full six feet long and three or four inches in diameter. We met on this river quite a number of this spotted, warlike tribe. Christ advised his early preachers to take lessons from serpents: "Be wise as serpents;" and we think, if the nations of earth would act upon some of the usages of rattlesnakes, we would be better off. They never fight except in self-defence, and then give fair warning.

We pushed on our way to San Patricio as rapidly as possible, in order to meet a body of surveyors that we knew were on their way to this point from Victoria. Arrived at the place in good time, and after hearty salutations were given and returned between us and the surveyors, and before we entered upon our work, we examined, with some interest, the relics of the little town once occupied by an Irish colony. No one had lived there since the campaign of 1836, when the Mexicans invaded the country, and drove all the colonists east.

Corpus Christi was the place where we agreed to pitch our camp; a name simply given to a locality on our south-western coast at the mouth of the Nueces River. We saw no indication of any former settlement at this place, but were informed, by an intelligent Irishman accompanying the surveyors, that this was the point at which the colony at San Patricio procured their supplies. We arrived at this point about sunset, and in consequence of the great amount of Indian signs, guards were promptly detailed and but little sleeping done.

The next morning the beginning corner was established, right on the bay, and the work went forward. The land on which the city now stands was taken up by our party. Wonderful indeed are the changes of these thirty-four years. While the work was going on every man's gun hung by a strap at his side. The surveyors numbered eight. My company numbered four. Our number was twelve, — in the midst of a country occupied by the most hostile Indians, some of them known to be cannibals, — and the nearest assistance, in the event of an attack upon us, was full ninety miles away. In view of this, every night, my company of four after dark went back into some secluded place, from two to three miles, so that the Indians could not find us, even if we were seen through the day. The surveyors were reckless men, and refused to follow the example, but camped every night where the work ceased. Which was the wiser policy will soon appear.

Uncle Matthew Burnett, one of my company, and I, determined that no Indian in the west had a horse that could run as fast and as long as ours. As our services were not specially needed with the surveyors, we agreed upon a stroll some eight or ten miles west, in order to examine further the Indian and wild-horse range. We travelled leisurely long the direction of a small creek, some six or seven miles, over a rough hog-wallow prairie, and were suddenly startled by a clear, shrill yell, right behind us. Riding hastily to the top of a little eminence close by, and turning our eyes back, we saw, between us and the camp, twelve Indians mounted on horseback, and clad in the habiliments of war. Observing that we were upon the lookout, they halted in full view, with their Mexican spears glistening in the sunlight. The nearest point of timber was four miles. The point of timber, the position of the Indians down the creek from us, over a mile off, and our position, formed a triangle. We knew enough of Indians to know that our safety depended on reaching the timber. They stood perfectly still, waiting to see what course we would take. Knowing that we were cut off from our company, and feeling confident that they could reach the timber first, they considered us a sure prize; and, indeed, it looked very much like it. Two against twelve, certainly, was a great odds on an open field. They yelled, and we yelled; but neither party moved for some time. They were one mile and a half nearer the timber than we were. Uncle Matthew and I had sufficient time to hold a council of war. Our plan was all laid. Every time they screamed the war-whoop we replied. My lungs were now apparently sound, and seeing what was before me, I straightened up in my stirrups and tried to feel that I was about twenty-one years old. We both had money belted round us, but in our war-council we decided that the race was not for money, but for dear life, and that, as they were cannibals, if we were caught our flesh would be eaten. We felt confident that we could beat all their horses to the timber, except four, that appeared at a distance to be in good condition. Should these get before us, we agreed to fight our way through them to the timber. As we moved off in the direction indicated, they started for the same point. As I rode the faster horse, I remained in Uncle Matthew's rear. We held our horses up for fully half the distance, determined to put them to their full speed towards the close of the race. Every Indian was whipping his pony with all his might; and every time they yelled we answered. When I was a boy it was a great relief to whistle when alone in the darkness, to keep my courage up, and this yelling answered about the same purpose. Before the race was more than half through, the Indians were scattered. The four good horses had distanced the others from two to four hundred yards, and as we rode up the hill out of the hog-wallow land, and on the half-way ground, the race was certainly a doubtful one. We were rapidly approaching each other in the form of an inverted V. Here each party yelled for the last time. I ran right up by the side of my friend, and with the end of my lariat whipped his poor, wearied animal with all my might. The race was very close, but we passed out a little in advance of the foremost Indian, and on reaching the timber leaped to the ground and prepared to shoot the leaders. No sooner did we present our guns than the shields were thrown up. They threw themselves over on the opposite side of their horses with nothing but an arm and a leg exposed, and wheeled back out of the range of our guns. We might have shot one or two of the leaders, but remembering that they all would be upon us before we could reload, we reserved our fire. By this time all the Indians were up, but, as they were afraid of our guns, they did not come up near enough to reach us with their arrows. Every time they made advances, we presented our guns and they fell back. Knowing that the Indians could speak the Spanish language, I addressed them in that language with hard names, calling them dogs and cowards. I dared them to advance on us, knowing that an Indian trembles in the presence of a resolute spirit. Being exceedingly anxious to get rid of them, I mounted my horse and galloped a few paces into the timber, and beckoned and called, part in English and part in Spanish, as though I was ordering a company of men out of the woods to come out and charge upon the Indians. By these demonstrations they were made to believe that assistance was close at hand, and they ran away in haste. We were, of course, greatly relieved.

The Indians went south, and as soon as they were at a proper distance, we made our way in a north-easterly direction to our camp. My mind was not at ease. In the midst of my danger, I had made a wilful misrepresentation to the Indians, making them believe there were men at hand, when there were none within six miles of the place. "Lie not one to another" is a plain command, and I was without question guilty of deliberate falsehood. My comrades then and afterwards made themselves merry at the recital of this scene; but the falsehood mixed with it has always cast a gloom over my mind. Our lives, however, were preserved, and I hope, in answer to prayer, God has forgiven me for any sin I committed while passing through this fearful trial. Nothing could have induced me to undertake that race again.

We now felt the necessity of great vigilance in guarding our surveyors while at work, and our camps at night. Provisions were getting scarce, except wild meat; life was in jeopardy every hour; and I would gladly, since my narrow escape, have given up all my interest in the wild western lands, to have been at home. But as my lot was cast in Texas, and as Texans in that day had little respect for preachers who gave any signs of cowardice; and as I desired to return home in "good report among them that are without," that my ministry might not be hindered, I contented myself the best I could, and hurried on the surveyors as rapidly as possible.

Man is a strange compound, and often knows but little about himself. Several days passed away; Matthew Burnett and I had but little employment. We grew weary of camp confinement; and as we saw signs of Indians about, and still nobody hurt, we became a little reckless again, and agreed on another expedition, as it would be yet several days before we could start for home. Parties of Lipan and Tonkawa Indians, friendly tribes, frequently visited our camp and told us of the country farther west, of the salt lake, and other points of interest, that we were anxious to see. They told us of the Camanches and Cacrankaways, who were enemies to each other and enemies to everybody else. The country we wished to see was occupied by these last-named tribes, who fought wherever they met. In consequence of so much war, the Cacrankaways were reduced to about forty warriors.

Matthew Burnett and I, after necessary preparation, were soon off for a trip of several days. Our course, by the compass, was a little south of west. We saw more wild mustang horses and wild game of every description than we had time to number. After travelling about forty miles, we found plenty of fresh water and good grass for our weary, thirsty horses, and struck our camp. No Indians had been seen, and the night was passed without interruption. The next morning, as our meat was out, we determined to kill a beef the first opportunity, as there were plenty of wild cattle in the range. We soon saw the remains of four beeves killed by the Indians. About noon we saw some cattle feeding at a distance, and taking advantage of a small bunch of timber, cautiously made our way to them. When close enough to shoot a fine beef, we saw a horse coming from the opposite direction straight towards the same beef, and after watching a moment, we saw an Indian behind the horse and driving him along. Discovering us, the Indian instantly fell to the ground, and strung his bow. This frightened the beef away, and soon the Indian was on his horse in plain view, about eighty yards distant. My friend Burnett raised his gun to shoot, but I insisted that his life be spared. We were in no danger, could not plead self-defence, and in the commission of a deliberate murder I feared the judgments of God. The Indian rode off, and as we rode along parallel with his course we commenced conversation in broken Spanish. I told him we were Americans, and his friends, pulled my cap off and put it upon the muzzle of my gun, showed him the spring dagger by the side of the barrel, but did not approach any nearer to him. Friend Burnett still insisted on shooting him; but I protested, and continued a friendly conversation. He was a young Cacrankaway, and pointed us to the smoke of their camp, plainly in view, but some distance away. He urged us to go with him to the camp and get acquainted with his chief. We had, however, seen enough, — aware of the fact that they knew the locality of our camp, and as we were so far away they would naturally suppose our company was scattered. With these facts before them, we were of the opinion that they would hasten to camp, hoping to surprise and capture it. No time was lost. We were between forty and fifty miles from our camp, and knowing the caracter of our enemy, it was necessary for us to reach the surveyors by the coming morning early, or in all probability it would be too late. We rode hard during the entire evening, and as much of the night as the horses could bear. We kept off of our former track, lest we might come in contact with the Indians, and travelled ten or fifteen miles farther, we supposed, than when we came out. About daylight, on the third morning after we started on the expedition, we reached the camp of our boys, then, as usual, about four miles from the surveyors. Preparations were made as soon as possible, and when we reached the main camp, we found Buchanan and his men surrounded by the Indians, their horses, guns and everything in the enemy's hands. The Indians had out-travelled us, knowing the country and taking a nearer route. The chief was on Buchanan's horse, and his forty warriors stood defiantly round the camp. Not a man as yet was hurt.

Four of us stood off about eighty yards, with gun in hand, and proposed a conference. I felt confident they were the same Indians whose smoke we were pointed to the day before, full forty-five miles towards the west. Accordingly I assumed command, and ordered Buchanan, then a prisoner, not to answer a question until it was first submitted by the interpreter to me. Under the order he assured the chief that we were his friends. He inquired of Buchanan, if any of his company had been out hunting the day before; the number and color of their horses, and character of their clothing. These questions were satisfactorily answered, and the chief said, "If these are the men, you are friends, or they would have killed my boy." We called for the boy to make the examination. He started to us with his bow, but we made him throw it down; and as soon as he recognized us he ran up smiling and shook hands with us, apparently as glad as if he had met relatives. We then ventured up closer, keeping our guns in readiness. Peace was soon made. Horses, guns, blankets, and everything was given up, and a treaty was made. I thank God yet that my motto ever was, even among Indians, not to kill except in self-defence.

We agreed to give the chief a letter to President Sam. Houston, then in the city of Houston, asking him to recognize our treaty. He left that day with three of his warriors, and one of our men to accompany him.

We remained a few days longer, got our lands surveyed and field notes written, and when we reached Victoria, met the chief on his return. Sam. Houston had signed the treaty, and, complying with the Indian custom, had made a number of presents. The chief left us near Corpus Christi, almost naked. Now he stood before us, full six feet and four inches high, weighing two hundred pounds, wearing a two-story silk hat, a fine broadcloth suit, and a fine pair of military boots, with a sword hanging at his side. He at once recognized our company, and ran to shake hands, but on approaching me took me in his arms. This was the first and only Indian that ever hugged me. I have no information that this treaty was ever violated. A profound impression was made by this incident, that every spirit begets its like. Kindness and mercy shown to an Indian boy resulted in saving the lives of our surveyors, and effecting a treaty whereby doubtless many other lives were preserved. God's word was verified. Coals of fire had been heaped on an enemy's head, and the animosity of a savage was consumed. The course that many of the early settlers of this country pursued, killing every lone Indian that was cut off from his company, was a great outrage, and certainly condemned by the word of God.

My principal object in this chapter has been to give the reader of this simple narrative the true state of the country in 1838, from the Brazos River west. The dangers that lay in our path were in the path of every man who ventured abroad. After an absence of two months I was again at home in the town of Washington, in the bosom of a happy family, who had heard nothing from me during my stay. At the time of this writing news can be sent by mail, and flashed over the wires; but then there was no communication.

CHAPTER IX.

GREAT ENCOURAGEMENT. — 1839.

DURING the summer of 1838, we had some encouragement of a religious character. In the fall gloom hung heavily over all the land. The government was unsettled. Sam Houston's term of office, as president of the Republic, expired on the second Monday in December, 1838, and according to the constitution he was ineligible to office for another term. He who had so long been as it were the very soul of Texas was about to retire, and there was a general feeling of anxiety as to the result of another administration. While there was no large army gathered anywhere, Indians, Mexicans, and Americans were on the war-path in amost every direction, in small companies. The house of worship commenced by the little handful of Baptists at Washington was a failure; most of the members were moving away, and the church was dissolved.

With a sad heart I determined upon another location, and early in the winter of 1838 moved to Lagrange, on the Colorado River. This was at that time a very small place. Above Lagrange, some six or eight miles, near where the Plum Grove Baptist meeting-house now stands, we pitched our camp.

The first Saturday evening after my arrival, brother Wm. Scallorn rode up to my camp and inquired if my name was Morrell — if I was a cripple — and if I was from the western district of Tennessee, all of which being answered in the affirmative, my consent was obtained to preach at his house the next morning. The appointment was circulated, and the few people then living in the neighborhood were assembled together. Before the sermon was closed the preacher plainly saw that there was quite an interest in the little congregation, and a regular monthly appointment was announced. At each successive appointment the interest increased, until the power of God was manifest in a precious revival, — the first I had enjoyed for a long while. During these meetings held in the spring of 1839, Elders Wm. T. and J. V. Wright, twin brothers, who have since preached in Texas with such power, were both convicted of sin, and subsequently converted and baptized.

On a visit to the town of Lagrange, in February, 1839, I heard of the Honorable R. E. B. Baylor, formerly a congressman from Alabama. A letter was handed to me which showed that he had professed religion, joined the Baptist church, and had been exercising his gift with great promise; so much so that the church had licensed him to preach. In this letter it was further stated that he had gone to Texas, and that the brethren were greatly exercised about his welfare as a preacher; that in consequence of his distinguished political attainments, and the inducements offered in a new country to seek promotion, fears were expressed lest he might not be active in his ministry, as there were no religious organizations to throw the mantle of protection about him. As he was then in town, I sought his acquaintance at once, and invited him to fill my appointment the following day. He declined to do this, but agreed to attend and aid me in the service. As preaching was at that time a novelty in Lagrange, the people all came out. After the sermon Brother Baylor closed the service with a very happy exhortation. He announced in the very outset that "there is a reality in religion and the Scriptures are true." This great thought was brought to bear with such power that it was easily seen that he was himself once an infidel. He contended, not only from the Scriptures, but from experience, that religion was a reality and ought not to be deferred. Here arose a bright star from the East, thirty-three years ago, that afterwards appeared with brilliancy in the Texas Baptist galaxy; and though it is now dim with age and infirmity, it still shines in our general convocations, and often reminds us of valuable service rendered in organizing the Baptist element of the "Empire State." I went home from this meeting greatly strengthened, blessing God by the way for this valuable aid in time of great need.

Indian depredations were carried on at such a fearful rate in 1839, stealing horses, killing and scalping the citizens, and sometimes carrying off boys and girls of twelve years old and under, that the citizens met together, and by resolutions declared themselves *minute men,* ready at a moment's notice, day or night, at their own charges, to go out in the common defence.

The reader doubtless remembers the little girl of poor Harvey, captured and sold into Mexico, and afterwards recovered. John McClellan, a lad of eight or ten years, after witnessing the killing and scalping of some of his relatives, was taken by the Indians and kept a number of years high up on Red River. In 1846, under a treaty with the Indians, he was obtained and conveyed to his father on the Brazos.

The lad had grown to be a man, had entirely forgotten his mother-tongue, but told by signs how the killing and scalping were done at the time of his capture. He was a perfect wild man of the woods. It was with great difficulty he could be kept from running off towards the Indian territory. Finally he agreed to stay one moon, and under a combination of young ladies, who visited and encouraged him, his affections were gained, and he became civilized. He afterwards married, and only a year or two ago I had information that he was living near Waco.

The Indians would come down on these expeditions about full moon, get up their booty if possible in the early part of the night, and travel by moonlight till day. During the day, while they rested, spies were kept in bunches of timber on the high prairies. The finances of the government were so weak, and the few soldiers so scattered, that they could seldom be gotten together in time to overtake the Indians. Thus our minute men were compelled at their own expense to keep guns, ammunition and war hoses, or allow these red men in most instances to go unpunished.

Our meetings at Plum Grove were continued. The instrumentalities were feeble, but God "out of weakness brought forth strength," and there were a few professions of religion. We visited the little organization at old brother Joseph Burleson's, twelve miles above Plum Grove. As Elder Abner Smith, their pastor, was paralyzed and helpless at the time, by request of the church I baptized sister Dancer into the fellowship of the church, who had professed conversion at the Plum Grove meeting. This was done about the first of March, 1839, in the Colorado River, some eighteen or twenty miles above Lagrange, and was my first baptism in the State. This was the first baptism that I have any account of west of the Trinity.

An announcement was made that several persons would be baptized into the newly constituted church at Plum Grove, at the next meeting, two weeks off. The time arrived, and we found persons present from the neighborhood of Columbus, forty miles down the Colorado River, and others had come from the settlements above as high as forty miles. Men living eighty miles apart took each other kindly by the hand at a little monthly Baptist meeting, their hearts bound together, not simply by the bonds that united frontier men so closely, but united by that spiritual union that God ordained should exist in Christ, before the sons of God shouted at creation's morn. After the sermon, and the ordinary preliminaries of a Baptist conference on Saturday, nine candidates came forward and gave the reasons of their hope. I was again greatly encouraged by the presence of brother R. E. B. Baylor, from Lagrange. Sunday morning, at ten o'clock, we met at the water, and after a short discourse on the subjects and action of baptism, nine converts testified their belief in the burial and resurrection of Jesus, by allowing themselves to be buried in and raised out of the water of the Colorado River. On retiring to a small house, with an arbor of brush built in front of it for the occasion, brother Baylor, in his usual happy manner, preached a most excellent sermon. Regular monthly meetings were held during the spring, with sixteen additions in all to the little church. This embraced about all the material then in the neighborhood, except the twin brothers, Wm. and J. V. Wright, who carried their convictions along, and were subsequently baptized.

This season of refreshing from the presence of the Lord greatly revived the drooping spirit of the way-worn traveller, and as we surrounded the table, with brother Baylor and these dear brethren, and ate the bread and drank the wine, my poor soul blessed God in faith that the wilderness of the Colorado would blossom as the rose, and that the solitary places along this fertile valley would one day be made glad. The Plum Grove Baptist church still lives; is the mother of the churches of the Colorado valley; and in her age is still blessed with a large membership and practical godliness.

CHAPTER X.

WHAT SHALL WE DO? — 1839.

WHILE the little band of Baptists was being increased and strengthened, the clouds of war, east, north and west, hung heavily over the land. Gen. Edward Burleson was the leading Indian fighter of the west, with Jack Hays, Ben and Henry McCollough, Caldwell, and others of like spirit, all in readiness at any hour to engage in the most daring expeditions. The spring and summer of 1839 furnished ample opportunities.

For about fourteen years the Cherokee Indians had held undisputed possession of Eastern Texas, north of Nacogdoches. Being an agricultural people, they had previously given no trouble. Cordova and other Mexican emissaries succeeded in breaking up the amicable relations between the Texans and Cherokees, and now Gen. Rusk in the east, in command of some five hundred men, was fighting and driving the enemy north. Bowles, the Cherokee chief, was killed in one of these engagements, and the Indians were driven off, of course greatly exasperated, and determined to avail themselves of every opportunity in future to avenge the blood of their chief and fallen warriors.

The tide of immigration continued to pour into the country. The small crops, made under disadvantages the previous year, could not meet the demands of the increasing population; provisions were scarce and high, and there was but little money in the country. We were hard pressed for subsistence, and the last peck of corn in my house was divided with the widow and orphan at Lagrange. I was then a citizen of that place.

News had for several days been circulated that a wagon-load of flour was due from Houston. As we sat eating about the last bread on hand, and a little anxious as to where the next would come from, my little son came running in from town, and stated that the wagon had come, and that the people were around it like a flock of blackbirds. The scanty meal was left unfinished, and in company with the lad I was soon at the wagon. The wagoner was rolling the barrels out, and whoever got his hand on the barrel first claimed it. The writer, full six feet two inches high, with arms in proportion, could reach about as far as any oher man present, and soon claimed flour by virtue of possession. The flour was rolled home, and soon the wagoner came for his money. No questions were asked about price when I took possession. The price was now asked, and stated to be fifty dollars. I was never so glad to get flour before; but was forcibly reminded that a few such purchases would give my pocket-book the swinney, — a disease right hard to be cured in those days by an *honest man.* Some of my property must be sold in order to meet my expenses, and with the hope of realizing more for it in San Antonio than at home, I made preparations for the trip.

From Lagrange to Gonzales fifty miles, there was but one house. From Gonzales to San Antonio there was no house. I knew the main Indian country, and determined to travel through this at night. Two Mississippians proposed to accompany me, and received my permission, provided they were willing to travel by my direction, and on my time. My son James, fourteen years of age, was to accompany me, making four in the company. We were to travel by the compass, and most of the way through the country without any road, taking observations by day and observing stars at night. With very definite understandings the journey was commenced.

We started in the evening, in time to make twelve miles by dark. My companions greatly admired the country passed over, and doubtless thought at first that I was very distant and unsociable. It was my habit, in all early and middle life, when starting on preaching tours, or on any other business of importance, to give myself up to reflection and prayer. On this occasion I felt more than usual the need of a special providence.

About dark we reached a small creek, and while my little son stood guard on his horse some distance back, we went under the hill and kindled up fire sufficient to make coffee, to keep us awake through the night. This was by no means the first time the lad had stood guard, and he understood his duty without any special directions. My companions inquired rather anxiously if we were in danger already. I informed them that we were in a savage country, and that my father, one of the sons of 1776, who went through the Revolution when a boy, and who acted as courier for General Marion, had taught me that eternal vigilance was the price of safety. After drinking coffee, I exchanged positions with my son, and when he was through, a large fire was built up to attract the attention of the Comanches, if there should happen to be any about, when we were gone. One of the company before leaving descended the bank of the creek to get water, and discovered the bones and skull of a human being, and a blanket close by. After a little examination, our decision was, that some traveller had passed this way, made a fire, and camped for the night. Indians had doubtless surprised, killed and scalped him, taking horse and all the effects. Some friends passing the same place had rolled up the body in a blanket, and buried it in the best manner possible with the means at hand and afterwards the wolves had disinterred the body and eaten the flesh, leaving the naked skeleton. We were all more than willing to leave that place, and travel all night, rather than camp in that vicinity.

We travelled till about one o'clock, and slept an hour, with one of our number on guard. The next morning we were at Gonzales for breakfast. Here was a community large enough to defend themselves against any ordinary force of Indians. After a day's rest and some refreshing sleep, late in the evening we crossed the Guadalupe, and as there were several Indian crossings ahead, we preferred another night-ride, especially as we had information that several men had recently been killed near these crossings of the road.

The next evening we were sixty-five miles from Gonzales and within ten miles of San Antonio, all safe, and out of the Indian range.

Here we came upon a large herd of sheep in charge of two masterly shepherd dogs, one in front and the other in rear of the flock. The dog in the rear barked, and rushing towards us in a furious manner, called us to a halt. The other dog rose and ran about four hundred hards to the top of the opposite hill, the whole flock running rapidly after him. As soon as the sheep stopped, the dog permitted us to pass, and ran rapidly to his station. His fidelity was tested the second time, with the same demonstrations. We camped near the place for the night, and watched these dogs at sunset bring the herd into the fold. This we were informed they did every day, without any assistance. After breakfast next morning, we went again to see the dogs carry the herd away, and inquired into the secret of this wonderful training. The puppies, we were informed, were taken from the mother before they recognized her, and had been raised by a favorite ewe. They were thus raised with the sheep, and naturally undertook the defence of the flock.

A pleasant ride over a lonely country brought us to the city of San Antonio. It was at this time in a dilapidated condition, but evidences were given of considerable expense and labor in the past. Several valuable stone buildings were seen and a large number of Mexican shanties.

The great object of interest at this time to us was the far-famed Alamo, an ancient Catholic mission house, and often used as a fortress. Within these walls my old Tennessee friend, David Crockett, had fallen, whose name is still familiar to every youth in the country, linked as it is to so many anecdotes. Here, three years before, on Sunday morning, March 6th, 1836, Travis, Bowie, and Crockett, with one hundred and eighty-five heroic defenders, had perished, after a desperate struggle with an overwhelming number of Mexicans under Santa Anna bearing a red flag. The incidents of that struggle, and the names of a large number of the slain, were then fresh in my memory, and produced emotions in my soul quite different from those now felt by the stranger who stands by the place, and hears the fearful incidents rehearsed.

Many things in that old city attracted my attention and furnished material for thought in after years. Catholicism then reigned without a rival. Before me were evidences on every hand of the blindest superstition. Since that time I have always supposed that the term priestcraft was a better appellation for this system than religion. If ignorance among the masses of a country will best subserve the interests of priests, then all the facilities for education are hindered. If intelligence ride into popular favor, in spite of priestcraft, then schools are christened in the name of religion, and a studious effort is put forth to gather all the children of the country within the folds of Catholic education. Strange that Baptists and Protestants patronize such institutions. Observations of a half century lead me to prophesy that before another half century has passed our children will regret the folly of their parents.

Two days were spent in the city, the business for which we had gone was attended to, and we were homeward bound. The Mississippians returned their thanks for valuable lessons relative to travelling through a savage country, and pursued their way to parts beyond, leaving me and my little son to travel back entirely alone. A different route was chosen on the return, and as we approached the Guadalupe River, an occasional volume of smoke on the east and west of the river gave the clearest evidence that the Comanches, in detached companies, were travelling in a south-easterly direction across the country. We had to pass right across their track, in their rear, and for fear of stragglers discovering us, we kept a sharp watch. We passed their track undiscovered, but saw a pool of water, yet muddy, where they had watered their horses. Steering our course clear of the Comanches, late the same evening an incident occurred, that will be more interesting after a short notice of a distinguished Mexican chieftain.

Cordova, whose name we have mentioned before, had been in the east stirring up the Cherokees and other friendly tribes to hostile feelings and demonstrations towards the Texans. Seeing the dangers that beset him and his comrades, since General Rusk and his five hundred intrepid Texan rangers were on the war-path, he determined to make his escape across the country into Mexico. Between sixty and seventy Mexicans, Indians, and negroes accompanied him. They crossed the Trinity River near Pine Bluff, and, leaving Springfield to the left, crossed the Brazos near where Waco now stands; crossed the Colorado at about the present locality of Austin, and, without any molestation that we know of on this route, were aiming to cross the Guadalupe a little below Seguin. Colonel Ed. Burleson, ever on the alert in the west, got news of him, and, with about eighty men under his command, encountered Cordova about three miles below Seguin, on the east side of the Guadalupe.

My little son and I had crossed the river, and fortunately had fallen in company with two citizens of the Colorado valley. We stopped to make a little coffee, to keep us awake on the night-ride to Gonzales, feeling quite grateful for our escape from the Comanches during the day, having been some time from home, and not knowing that Cordova or Burleson were either within a hundred miles of the place. While resting and enjoying the coffee, a single gun was fired about a half mile east of us. Supposing it was some citizen from Seguin hunting, we paid but little attention. But in a moment two or three guns fired; then a half dozen; then by platoons the firing increased, and was coming steadily nearer to our camp. By the time we could get our baggage up, and ride two hundred yards to the top of a little hill, Colonel Burleson had driven Cordova, in a running fight, right over our camp fire. We watched till the retreating party made their escape to the timber below; and, when the firing ceased, not knowing who the parties were, we made our way on towards Gonzales, in the night, as rapidly as we could. At Gonzales we learned the particulars of the fight, and that Burleson had killed eighteen of Cordova's men, without losing a man in his command. Although they passed within a few hundred yards of us, it was getting dark, and we could not decide who they were.

Having travelled till late in the night, worn out by the excitement and fatigue of the day past, we stopped to rest on an elevated, open, mesquite prairie, with only a few small trees near us. The moon was shining beautifully. Each man, by turns, was to stand an hour. Guns were all examined and carefully laid at our sides. Horses were staked, and before retiring I reminded the watch to call me, if he heard an owl, a crow, or a wolf. By this time I had learned that Indians imitated all these, and sometimes in this way surprised the traveller, besides using these sounds as signals for companions to assemble. The first watch had not expired before I was called up and notified that an owl had been heard. Quietly waking all up, I raised my head above the grass and watched, in the direction of our horses. I soon discovered three mounted Indians approaching my favorite horse, on which I had outrun the twelve Indians in the west, the year before. Crawling out near the horse, and determined to sacrifice life before they should have him, I waited till the Indians were within about thirty feet of him, and suddenly rising to my feet, with my gun to my face and in plain view, I gave the *Comanche yell.* As they wheeled to run away, the boys begged permission to shoot; but, as stated before, unless in self-defence, I was opposed to the policy and principle. As this was not a good place to rest, we mounted our horses and rode on to Gonzales.

In the midst of all these troubles on my arrival at home, the education of my children was pressing with great weight on my mind. Three years and a half were gone since my move to the State, and no schools yet, worthy of the name. My hopes of educating them in Texas had all vanished. The decision must be made, either to return them to the land of their birth, or allow them to grow up in ignorance, amid the wilds of the wilderness, and become "hewers of wood and drawers of water" for the intelligent population that in their generation would cover all these rich lands of the west. The three children remaining at home, two sons, aged respectively seventeen and fifteen and a little daughter, nine, were each put on a horse, and we started for Houston, intending to take water. Reaching that point, we found the yellow fever raging to a fearful extent. Of course it would not do to expose the children to this fearful disease, and overcoming the temptation to return, I renewed the supply of provisions, and undertook the long journey on horseback, knowing that in some places it was forty miles between houses. On reaching Hinds County, Mississippi, I found brother S. S. Latimore engaged in a protracted meeting. Here I stopped to rest myself and children a few days and enjoy the meeting. Instead of carrying them on to Tennessee, I left my two sons under brother S. S. Latimore, the president of the school at Middleton, and my little daughter at Lexington.

As my children were left behind, rather than make the long, tedious journey on horseback alone, I determined to face the yellow fever, and trust in God for my deliverance. The epidemic was raging fearfully at New Orleans and Galveston on my return. My only child, except the three left behind, was a married daughter, who with her husband had remained in Houston. On my arrival there, the fever was subsiding for want of material. Both son-in-law and daughter were just recovering from the fever.

The only income I had was derived from three wagons that hauled freight from Houston to the interior. These were driven by hired teamsters. The wagons had been started with freight to Austin, and my son-in-law was to follow on and see the freight delivered. His sickness had prevented, and a letter was received, on my arrival in Houston, that the teamsters had drawn six hundred dollars, sold a portion of the teams, and were trying to sell the remainder. I rode home as soon as I could, and tarrying only one night, made my way on to Austin, and found that, according to the laws of the land then, I could not recover the six hundred dollars advanced to the teamsters. Returning by way of Bastrop, I recovered the remnants of my teams. The heavy losses and my expenses in the education of my children brought upon me a heavy burden at this crisis. It was indeed grievous to part with the church at Plum Grove, where such precious meetings had been enjoyed and the first fruits of my Texas ministry had been realized. But for good reasons our location must be changed, and a permanent home provided, if possible.

Hoping to get out of the Indian range and make a permanent settlement, I traded my land for lands on the Guadalupe, thirty miles above Victoria.

CHAPTER XI.

WAR. — 1839.

LONG as memory holds her seat will the early settlers of Texas remember the events of 1839. Harassed by war on every hand, the unsettled state of society made our circumstances almost beyond endurance. Our currency was almost worthless, and the Republic without credit abroad. Gradually sinking in value, the money finally, in the year 1840, fell to the small sum of fourteen cents on the dollar.

The Indians grew more hostile and troublesome. Our minute men had been called on so often to drive the Indians from the settlements, that they determined finally to follow them up to their hiding-place, if possible, and punish them there, hoping this would have good effect. Accordingly Colonel John H. Moore, of Fayette County, got up an expedition and travelled far up above Austin, making his way up in the night with the aid of friendly Indians, and found the enemy camped in a bend of the river. The bend resembled a horseshoe, with a high bluff on the opposite side of the river. They were surrounded while asleep, and at daylight a destructive fire was opened upon them, demolishing the camp and killing large numbers. Many swam the river, and being shot while climbing the bluff on the opposite side, fell back into the water. A few attempted to escape by going down the river and crossing at a ford below. These were followed up and killed or captured. The citizens bore the expenses of the expedition.

The Indians suffered much during these years. as well as the Texans, and in February, 1840, a squad of Comanches went to the city of San Antonio and proposed to make a treaty of peace with Texas. They were promised that if they would return the captives they had stolen from Texas, peace would be granted. This they promised to do on the next full moon.

At the time appointed several chiefs and quite a number of the tribe made their way to San Antonio, leaving some of the captives behind. A difficulty occurred relative to the terms of the treaty. The chiefs were informed that they would be held as prisoners until the captives they had carried out of Texas were all brought back. Upon this the chiefs made an attack, and continued to fight until they were

killed. The punishment inflicted by Colonel Moore and his minute men, and the killing of these chiefs, aroused the Comanches to such fury, that preparations were made for vengeance upon a large scale on the white settlements towards the coast. Early in August, 1840, they swept down the country in very large numbers, and before the citizens of Victoria were aware of their approach surrounded the town. The citizens rallied together promptly an drove them away, carrying as they went large numbers of horses and cattle from the prairies. They went clear to the coast, and sacked and burned the little town of Linnville. Several persons were killed, and Mrs. Watts, a lady from Linnville, was carried off a prisoner, her husband having been killed in her presence.

My wagons had previously been loaded with lumber at Bastrop, which was safely deposited at the place for which I had traded on the Guadalupe, thirty miles above Victoria. On my return, between the Guadalupe and Lavacca Rivers, I saw clouds of smoke rise up and suddenly pass away, answered by corresponding signs in other directions. We passed with the wagons just in the rear and across the track of the Indians as they went down. From their trail I thought, and afterwards found I was correct, that there were four or five hundred. The trail was on the dividing ridge between Lavacca and Guadalupe Rivers. I trembled for the settlements below; for I knew this meant war on a larger scale than usual. About two miles after we passed this trail, we found a horse whipped and spurred till he could go no further. Just at this time, a herd of mustang horses, almost run to death, passed about one hundred yards behind our wagons, pursued by a body of twenty-five or thirty Indians. Seeing our guns and pistols, the Indians turned off and kept out of the range of our fire-arms. Above Austin they had attacked a wagon and thirteen of our men; and although they captured the wagon and killed twelve of the men, it had cost the savages so many lives that they did not care to come in contact with wagons at so early a date the second time. This we presume was, under the providence of God, the reason of our escape. They could have overpowered us in a very short time. This was doubtless the rear guard of the advancing, barbarous plunderers.

About a half mile from where we saw the mustangs, a party of stragglers had attacked two men. One of them, being shot, fell from his horse, and they, supposing him to be dead, left in pursuit of the other. They soon captured him and brought him back to where the first had fallen. Immediately after they overtook him, they cut off the soles of his feet, and made him walk barefooted on the rough grass back to where the attack was made, hoping, we suppose, after the cruel treatment was over, to get the scalps of both. On their arrival at the spot, the man whom they supposed to be dead had crawled to a neighboring thicket, badly wounded. Fearing to attack him, knowing that he had a gun, and was securely sheltered from their arrows, they took his companion's own gun and shot him dead, terribly mangling his body, in plain view. The man in the thicket saw my wagons pass near by, a few minutes afterwards, as he subsequently told me. My oxen were in fine condition, and being anxious to communicate this intelligence to Colonel Ed. Burleson and the citizens of the Colorado valley as early as possible, I drove thirty miles in twelve hours. I crossed the Indian trail at twelve o'clock in the day, and reached home, at Lagrange, at midnight. In view of the long race before me, I tried to sleep some, while a horse was being secured. At four o'clock in the morning I was in my saddle, intending to reach Colonel Ed. Burleson's at daylight, twelve miles off, on a borrowed horse, as I had no horse in condition for the trip.

The reader must pardon me for a little digression. An incident in the history of the owner of the horse that I had borrowed for this trip will illustrate the confidence felt that no Comanche in the range would overtake me in a fair race. This confidence in those days rendered great relief. Captain Dick Chisholm, the owner, lived a year or two before down the Guadalupe, between Victoria and Gonzales. Chisholm weighed two hundred pounds, and ventured out on business one day, between the Guadalupe and Lavacca Rivers, in company with a Texan, who rode a horse trained to run from Indians or after them. The captain was riding a slow, untrained horse. Soon twenty-seven Comanches were on their track. When Chisholm saw that he must be overtaken, he bid the Texan flee to Gonzales for his life, and tell the news. His slow pony fell down on the hog-wallow land, and the cannibal band stood around him. The Indians represented different tribes, and were compelled to talk Spanish, which he understood. The chief propounded a number of questions, but he refused an answer. In the mean time, the savages discussed the division of the fat man, after he should be killed, pointing out and marking off the desirable parts. He finally determined to threaten them and try to induce them to release him. The people at Gonzales, he informed them, since the two men were killed over this way a few days ago, have determined, if another man is killed, to follow the murderers as long as water runs and trees grow, or until they are exterminated. "Kill me now if you dare, and as soon as my man gets to Gonzales, a large body of men will come here, on horses as fast as his, and they will very soon run down the last horse you have; and they will kill the last one of you, and give your flesh to the buzzards. Kill me and eat me as quick as you please." The chief's eyes flashed with surprise; fear took hold of him; the horse and blanket and gun were given up; and after an assurance that they and he would ever afterwards remain friends, Captain Dick Chisholm made his way to Gonzales, determined never to ride a slow pony in Indian range again. The sorrel, ball-faced horse was purchased, — full seventeen hands high, seven years of age, and made well for a long race, — at two hundred and fifty dollars. My friend cheerfully tendered me this horse for the hazardous expedition.

The sun was just rising as I reached Colonel Burleson's house. The story was rapidly rehearsed. His war-horse was ordered at once. Just before mounting he pointed my attention to his saddle, wearing the marks of bullets, — one on the inside of the horn, one on the outside of the horn, and one on the back part of the tree. "All these," said he, "were made when I was in the saddle." His horse was killed under him at the battle of San Jacinto. By the time we were mounted, a man was in sight, his horse running rapidly, and a paper in his hand, fluttering in the breeze. The expressman presented the paper, which read about as follows:—

"GENERAL: The Indians have sacked and burned the town of Linnville; carried off several prisoners. We made a draw-fight with them at Casa Blanca, — could not stop them; — we want to fight them before they get to the mountains. We have sent expressmen up the Guadalupe.

"(Signed) Ben. McCulloch."

We made our way up the Colorado valley as rapidly as we could to Bastrop, notifying everybody as we went. Here Colonel Burleson called a council, and it was agreed that the Indians should be intercepted on their retreat at Good's on Plum Creek, twenty-seven miles below Austin. Colonel Burleson requested me to follow up the expressman to Austin, and urge the people to come forward promptly to the point designated. Here I rested at night, after a circuitous ride to Austin of about seventy miles. In the morning, rising early, we rode to the point designated, and found Colonel Burleson and his men had been gone about thirty minutes. Riding very rapidly, we came up with the Texan forces some two or three miles, as well as I remember, south-east of the present locality of Lockhart, and at the fork of Plum Creek. Colonel Burleson had been in communiation with the troops of the Guadalupe, and now Felix Huston, Jack Hays, Ben and Henry McCulloch, and others, were on the ground. General Felix Huston was in command, and preparations were being made for the fight, when I and the company from Austin rode up. The fight immediately opened, with about two hundred Texans, against what we supposed to be five hundred Indians.

The enemy was disposed to keep at a distance, and delay the fight, in order that the packed mules might be driven ahead with the spoils. During this delay several of their chiefs performed some daring feats. According to a previous understanding, our men waited for the Indians, in the retreat, to get beyond the timber, before the general charge was made. One of these daring chiefs attracted my attention specially. He was riding a very fine horse, held in by a fine American bridle, with a red ribbon eight or ten feet long tied to the tail of the horse. He was dressed in elegant style, from the goods stolen at Linnville, with a high-top silk hat, fine pair of boots and leather gloves, an elegant broadcloth coat, hind part before, with brass buttons shining brightly right up and down his back. When he first made his appearance he was carrying a large umbrella stretched. This Indian and others would charge towards us and shoot their arrows, then wheel and run away, doing no damage. This was done several times, in range of some of our guns. Soon the discovery was made that he wore a shield, and although our men took good aim the balls glanced. An old Texan, living on Lavacca, asked me to hold his horse, and getting as near the place where they wheeled as was safe, waited patiently till they came; and as the Indian checked his horse and the shield flew up, he fired and brought him to the ground. Several had fallen before, but without checking their demonstrations. Now, although several of them lost their lives in carrying him away, yet they did not cease their efforts till he was carried to the rear.

Their policy was now discovered, and Colonel Burleson, with his command on the right wing was ordered round the woods, and Colonel Caldwell, on the left, with his command, charged into the woods. Immediately they began howling like wolves, and there was a general stampede and vigorous pursuit. The weather was very dry, and the dust so thick that the parties could see each other but a short distance. Some fourteen or fifteen Indians were killed before the retreat, and a great many more were killed afterwards. Our men followed them some fifteen or eighteen miles.

Just as the retreat commenced, I heard the scream of a female voice, in a bunch of bushes close by. Approaching the spot, I discovered lady endeavoring to pull an arrow out that was lodged firmly in her breast. This proved to be Mrs. Watts, whose husband was killed at Linnville. Dr. Brown, of Gonzales, was at once summoned to the spot. Near by we soon discovered a white woman and a negro woman, both dead. These were all shot with arrows, when the howl was raised and the retreat commenced. While the doctor was approaching, I succeeded in loosing her hands from the arrow. The dress and flesh on each side of the arrow were cut, and an effort was made to extract it. The poor sufferer seized the doctor's hand, and screamed so violently that he desisted. A second effort was made with success. My blanket was spread upon the ground, and as she rested on this, with my saddle for a pillow, she was soon composed and rejoicing at her escape. Death would have been preferable to crossing the mountains with the savages. She had ridden a pack-mule all the way from the coast, and when they stopped she was required to read the stolen books for their amusement. I received many lettes from Mrs. Watts in after years, but never saw her again.

When we went into the fight there were present about two hundred men; but by night we supposed there were near five hundred. They continued to come in all the evening; many of them from a great distance. Men and boys of every variety of character composed that noisy crowd, that was busily engaged all night long talking of the transactions of the previous eventful days. Here were three Baptist preachers, — R. E. B. Baylor, T. W. Cox and the writer, all in the fight, with doctors, lawyers, merchants and farmers.

Glad indeed that the enemy was driven out, but weary and careworn, I made my way home, inquiring, How long shall these things be?

CHAPTER XII.

THE FIRST ASSOCIATION. — 1840.

DARK as were the days of 1840, God sent a little ray of light to shine upon the path of his servants, scattered in the wilderness. Although the Baptists were weak and few, they were slowly increasing, partly by experience and baptism, but mostly by immigration; and there were so many different opinions on doctrine, that a conference of the whole, to consider the common interests of our great Master, was a pressing necessity. This necessity had profoundly impressed the mind of the writer during the whole of 1839 and the early part of 1840.

In June, 1840, a small company of breathren, with four preachers, R. E. B. Baylor, T. W. Cox, A. Smith and A. Dancer, met in the town of Independence, Washington County, to form a Baptist association. Two of these preachers were missionary, and two anti-missionary. As the body could not harmonize, another appointment was made for October, at the town of Travis, Austin County.

Accordingly, on the eighth of October, 1840, messengers from three churches, — Independence, Washington County; Lagrange, Fayette County; and Travis, Austin County, — met at Travis, to organize what is now known as the "Union Baptist Association."

The church at Independence was organized in 1839, by Elders T. W. Cox and Spraggins, with eleven members. The church at Travis had been organized by Elders R. E. B. Baylor and T. W. Cox, with seven members; the church at Lagrange, by Elders J. L. Davis and Cox, with nine members.

In the month of September previous, I had moved my family to the Guadalupe, expecting to return in time for the organization; but was taken sick, and neither I nor messengers from the Plum Grove church were present. The membership of this church was larger, in consequence of the recent revival, than either of the three represented.

At the first session of the association, there were present three ministers, Baylor, Cox, and Davis, with messengers representing forty-five Baptists. Under the all-seeing eye of God these brethren deliberated, and laid the foundation of this mother of Texas associations.

The second session of this body was held at Clear Creek, Fayette County, commencing the seventh of October, 1841, with messengers from nine churches, representing three hundred and eight-four members. At this session a resolution was adopted, recommending the formation of an "Educational Society." This was responded to by the organization on the spot of the "Texas Baptist Education Society." The moving leading spirit in the enterprise was Elder William M. Tryon, — a man of the first order of natural and acquired ability. A short time previous to this I met him, and spent a few days with him in his family and among his congregation. Here I was impressed with the fact that he was not an ordinary man. A rare combination of excellent qualities fitted him for the pastorate, and evidences were manifest that as a missionary he must succeed. Elder Tryon was born to be a leader, and when Union Association gave birth to the "Education Society," he took the child by the hand, and from that day till his death he was emphatically the leader in the cause of education among the Baptists of Texas.

He was born March 10, 1809, in the city of New York. A part of his early life was spent in Connecticut. For the benefit of his health he removed to Savannah, Georgia, in his nineteenth year. He was left without a father at nine years of age; but under the influence of a pious mother he became a Christian in early life. His education, that so eminently qualified him for his after work, was received at Mercer University, and in 1837 he was ordained by Elder Jesse Mercer and others. He served as pastor at Eufala and Wetumpka, Alabama, until January, 1841; at which time he moved to Texas, under an appointment as missionary of the Triennial Convention.

He first settled in Washington County, between the towns of Independence and Washington, and devoted all his powers to the cause of his Master in the surrounding country. As indicated before, he was the man for any field. As pastor, few could excel him, as missionary, he wielded an influence that gathered all classes, from the poor to the rich, and from the most illiterate to the most refined and cultivated, about him; and when he espoused the cause of education he was master of the field, and moved the great Baptist heart to rally around the infant institution at Independence, and labor industriously to provide means for the education of the rising ministry of Texas.

Brother Tyron was a man of medium size, with an erect, well-proportioned physical structure. His dark, penetrating eye, above which appeared a full, well-formed forehead, impressed every man who looked upon him, as to his intellectual powers. Warm-hearted and eminently social in all his bearings, with a clear understanding of human nature, he was guilty of but few improprieties, and was manifestly on all occasions a Christian gentleman. His piety was deep and earnest, and while he was by nature a leader among his fellows, his great desire was to lead them all to God.

As an orator, my profound conviction is, that no preacher has ever lived in Texas who was his equal. He was well versed in the history and principles of the Baptists, and when his powers were brought to bear on this and kindred subjects, the charge, so often brought against us, of bigotry and ignorance, was hurled to the ground by this princely speaker. It was my fortune to hear him at many of our annual meetings, at his churches, and on missionary fields, and on all occasions he ever swayed the masses at will.

During the term of the Congress of the Republic that was held in the old town of Wahington, in 1843 and 1844, he served the body as chaplain, by consent of his churches, when the financial condition of the country was such that no remuneration was expected for his services.

In the winter of 1846 he moved to Houston, and took the pastoral oversight of the little church in that city. This was the last but crowning work of his life. Here was a small body of Baptists, without a house of worship; but, under the fervor of his ministry, the small congregation swelled to a large number, and the little church soon contained almost one hundred communicants. The body, under his ministry, put forth an organized activity and erected the present house of worship.

His career of usefulness in Texas was short. The last time the denomination was permitted to sit with him in council, was at the session of Union Association, held with the Houston church, in October, 1847. At that meeting he was elected and served as moderator. There were a few cases of yellow fever in the city at the time. After the adjournment of the body, and the fever was declared epidemic, he preferred to remain among the people whom he loved, and who loved him, notwithstanding the dangers that surrounded him. On the sixteenth day of November, 1847, he died, after much suffering, from a violent attack of yellow fever. His remains were deposited close by the church edifice that he had labored so hard to erect.

The deepest gloom hung over the entire denomination on the announcement of his death. Although dead almost a quarter of a century, he lives, and wields an influence over the people of Houston. His labors of love are still fresh in the memory of the brethren at Washington and Providence, and every student that has been educated in Baylor University owes him a debt of gratitude.

In this connection I will record the fact that Elder James A. Huckins rendered valuable service in building up our institutions of learning. He was a man of education himself, and on all occasions pressed its claims upon the masses; but especially he urged the importance of a well-trained mind upon those who intended to devote their lives to the work of preaching the gospel. His name appears recorded among the founders of the "Education Society."

He came to Texas under the appointment of the same Board that sent brother Tryon, and commenced his ministry here in 1840. Soon after his arrival, the same year, he organized the church at Galveston, and the following year the one at Houston, and preached for some time alternately in these cities.

I met him first on his field, in the spring of 1841, and held my second interview with him at the second session of the association.

He was a close thinker, and from the pulpit presented his thoughts in the clearest manner; always exhibiting the fact, that he was a profound scholar and student still. He preached almost exclusively from manuscript, and was seldom caught on any occasion without something appropriate in this form. It has always been to me a matter of astonishment that able ministers will uniformly stick to their manuscripts, when such good opportunities are so frequently given to stir the masses with the popular extemporaneous style of delivery. When on one occasion, at a general meeting at Independence, the appointee failed, brother Huckins, under an earnest protest, was driven to the stand. He urged that he had "neither long notes nor short ones," but no excuse would satisfy the brethren. His text was in Exodus xv. 11, — "Who is like unto thee, O Lord, among the Gods? Who is like unto thee, glorious in holiness, fearful in praises, doing wonders?" For the space of an hour held the audience spellbound, by the force of his clear, burning thoughts. At the close of that sermon, a number of us went forward and gave him permission to preach as often as he wished, even if his notes should be forgotten.

His power was felt by the denomination when he took the field as agent for Baylor University. Much that has been done, in endowment and building, was effected under his agency. The dates connected with his work are not before me.

Some time previous to the war he moved to Charleston, South Carolina. On that field he ranked high in the estimation of his brethren. During the war he ministered to the temporal and spiritual wants of the soldiers around that city, with a devotion closely allied to a martyr spirit. He fell at his post of duty before the contest closed, and rests in the bosom of God, where bitterness and strife never enter.

The first officers of this society were, Elder R. E. B. Baylor, President; S. P. Anderws, Recording Secretary; Elder William M. Tryon, Corresponding Secretary; Brother Collins, Treasurer. The board of Managers consisted of Elder James Huckins, Elder Z. N. Morrell, and brethren J. L. Farquahar, Gail Borden, Stephen Williams, William H. Ewing and J. S. Lester. Believing this to be one of the most important steps ever taken in the history of the Baptists of Texas, and in view of the influence it exerted over the entire denomination, and especially the rising ministry of the State, I have given the names of its founders. You may expect, further on in this work, a notice of the development of this child.

One incident occurred early in this, the second session of the body, worthy of notice. Brother R. E. B. Baylor, for some cause then unknown to me, had failed to give me the happy greeting and cordial saluation that he had on all former occasions manifested. Whatever might be said of him otherwise, my former association with him had led me to know that he was a cordial man; and I knew these demonstrations had sprung from something that he considered of grave importance. I was deeply wounded, and the pride of my nature forced me in return to treat him with marked indifference. He availed himself of the first opportunity after the association was organized "to make some remarks relative to Brother Z. N. Morrell," stating, further, that he "felt under obligation to make some acknowledgments." He stated openly the unkind demonstrations we had made toward each other, and explained the reason of his course, as the offending party. Brother Huckins a short time before this, as the pastor at Houston, had invited a man by the name of Merrill, a bad character, from the communion table. The name had been given as Morrell, and the impression made in the interior was that Z. N. Morrell was the man. Seeing that brother Huckins on his arrival had recognized the brother against whom the accusation had been made, he now knew the charges were false and hoped the explanation would be satisfactory. I immediately, with the permission of the moderator, arose and cordially accepted the explanation and the spirit with which it was made, as more than satisfactory.

It is true that in those days we did some things in our general meetings without as much system as is now observed, but here was a course pursued, and a spirit manifest, worthy of imitation by the true friends of Christ, until the Master comes the second time. If brethren would explain their mistakes promptly and, as publicly as they are made, give good reasons, it would heal up many difficulties and inspire confidence in place of alienation. There is also no greater evidence of godliness, than for men of superior attainments to "condescend to men of low estate," and treat with courtesy their brethren of inferior and less honorable positions in life. Here was a man of distinction, and so recognized, an able advocate at the bar, and fresh from the Congress of the United States, who had taken upon himself the office of the Christian ministry. On the other hand was the writer, of humble birth and limited education, known in Tennessee as the "cane brake preacher," and in Texas as a "brier cutter and Indian fighter;" and yet, notwithstanding the differences in other respects, we were equals in office as ministers of Christ. This simple incident, I will here record, wrote attachment upon my heart for brother Baylor that the revolutions of these thirty-one years have never wiped out. We have sometimes differed in opinions since, but whenever we have, my mind has reverted to the second session of Union Association, and a response always come up from my heart, "His head may be wrong, but his heart is right." The rising ministry of the country can take this for what it is worth, when brother Baylor and I, who are both now waiting at the sunset of life to cross over the river, have passed away.

Valuable work was done at this session of the body, looking to the organization of the churches, in order to beat back the heresies that were finding their way into the country. Not having the minutes of the first meeting at hand, I do not know who was the moderator. At this, the second session, Elder William M. Tryon acted as moderator, and William H. Ewing as clerk.

Owing to the dark hours of war, and the bereavements of brethren and sisters through the country, the frequent removal of families, and the scarcity of ministers, the minutes of the third session of the association show but a small increase of membership over the second. At this meeting were six ministers, and representatives of four hundred and forty-three Baptists. As an exhibition of the state of things among the churches, I here insert extracts from letters sent up by two of our churches, which were published with the minutes of this meeting: —

"Plum Grove Church, to the messengers composing the Union Baptist Association, when convened with Mount Gilead Church, Washington County: —

"Dear Brethren: In the year 1838, we first met together and held prayer--meetings. Soon thereafter we were visited by brother Z. N. Morrell. In 1839, our little church was organized by a presbytery, composed of brethren R. G. Green and Asael Dancer, adopting at the time of the organization the articles of the United Baptists of West Tennessee. During the year 1839 we were supplied with preaching by brethren Z. N. Morrell, A. Smith, and others. It was during this year that the ordinance of baptism was first administered in this part of the country. In August a candidate was immersed by brother Morrell, fifteen miles above this place, and soon thereafter fifteen received the ordinance and were united to our church. During 1840 and 1841 were were without a pastor, and unhappy difficulties occurred, which resulted in the dismissal of eight of our number. Our little church has suffered much during the late invasion. some of our beloved brethren and many of the dear congregation have fallen upon the field of battle, whilst others, and among them the son of brother Morrell, our pastor, are being carried as captives into the enemy's land. Truly, dark and thick clouds envelop us. Brethren, do not the calamities with which we are surrounded call, and call loudly, upon us to invoke the aid of Almighty Power? Let us go speedily, and pray before the Lord. Yea, let us approach boldly a throne of grace, and petitionfor help in this the time of *our* and *our country's* need.

From the church at Gonzales: —

"Dear Brethren: The history of our denomination in this portion of our country is as follows: Brother Z. N. Morrell is the first Baptist that ever preached in our country. He commenced laboring with us in 1840, and at the close of 1841 ten were collected together, who held letters from Baptist churches in the United

States. With these we constituted a church, adopting at the time of its organization the articles of faith held by the United Baptists of West Tennessee. Brother Morrell was called to the charge of our little church, and the regular monthly meetings and weekly prayer-meetings were kept up, until the time of the Mexican invasion last spring, at which time our pastor, and all others capable of bearing arms, left to repel the invading foe. Since then, until very recently, we have had no meetings for public worship. We have to lament that no additions have been made to our number since the time of our organization. Prospects last spring were encouraging, and some, we trust, were hopefully converted to God; but the unsettled state of the country was such as to prevent our troubling the beautiful streams of our country with the baptism of willing converts. Dear brethren, we are truly an afflicted people, but we rejoice that it is written, that though 'sorrow endureth for a night, joy cometh in the morning;'' that notwithstanding we may be slain by the savages, or by our enemies the Mexicans, we still have ''a hope, which is an anchor to the soul, both sure and steadfast, which entereth into that within the vail, whither Jesus, the forerunner, hath for us entered, an High Priest forever;' which hope cheers and supports us under our trials.''

These letters give fair samples of the trials and conflicts to which we were subject.

You will now remember that a Committee of Correspondence was appointed to the little church organized at Washington, in 1837. This correspondence resulted in bringing Elder Wm. M. Tryon and Elder James Huckins to Texas. These brethren afterwards served as missionaries in Texas, under the patronage of the ''Domestic Mission Board of the Southern Baptist Convention.'' The rest of us were nearly all missionaries in the Union Association, very much on the same principle that a majority of the soldiers of the Texas Republic served, at that time, at their own expense. The association was really in a formative state up to the sixth session, held with Mount Gilead Church, Washington County, when its present ''articles of faith'' were adopted. Of this troublesome period in our history I am always reminded when I see the name adopted — *Union.* Here were brethren from all the Southern States, and some from the North, with their pet plans and notions and opinions, — every man, almost, striving to get the articles of faith and plan of operations under which he had been trained as a Baptist, adopted, apparently, manifesting the belief that the Baptists of this particular State were the only orthodox people upon the face of the earth. The more thoughtful among the brethren went to every meeting of the body during this period, anxious lest good brethren, whose influence and co-operation were needed, would fly off at a tangent, and every means was adopted by them, without sacrificing vital principles, to hold all together. So earnestly did these brethren strive to hold together, that heresy was allowed to come in among us, and remain longer than strict compliance with the law of God permitted, as well be seen hereafter. None save those who lived through it, or through like ordeals elsewhere, can conceive of the many difficulties that beset us.

At the fifth session of the body, after dissatisfactions had been expressed at all the previous meetings after the first with the original articles of faith, a committee was appointed to revise and report to the sixth session. Pending the adoption of this report, and while Virginians, Georgians, South Carolinians, and representatives from other States, were urging the distinctive features of the old States, expecially with reference to the name the association should wear, — United Baptists, Regular Baptists, Missionary Baptists, all being presented and insisted upon, — one brother arose, and suggested that we were now Texans, and, as Baptists was sufficiently definite to distinguish us, insisted that the appellation of "Union Baptist Association" covered the ground. Elder A. Buffington, from Washington church, came to his relief, and stated that he thought the position well taken, and that "he had been constantly looking for some brother to offer a resolution to muzzle the dogs in persimmon time, because his ancestors from North Carolina had set the example." The articles of faith as they now stand, and the constitution, with some slight changes since, were at the sixth session adopted, and the association has maintained a remarkably harmonious career up to 1871.

The man who assisted in organizing the three churches that composed the Union Association at its organization, introduced the heresy, and was the cause of all the trouble on points of doctrine that the body ever had. This man was Elder T. W. Cox, from Alabama, — a man of eloquence and great natural ability. During the second session he preached on the subject of faith, and, departing from the doctrine plainly set forth in the New Testament, clearly taught the errors embraced in the system commonly known as Campbellism. The association being held with the church of which he was the pastor, he, at the close of the sermon, offered an opportunity for the reception of members, and was about passing decision upon their fitness for baptism, according to the approved plan of the Bethany School, when Elders Huckins and Tryon entered their public protest. Quite a debate ensued, and some confusion, but the reception of members was postponed. In a council held by the ministers present, it was agreed that charges ought to be preferred against him before his church. Church authority is the highest tribunal on earth, according to the New Testament, and the association could have nothing to do with the discipline of churches, except to advise. That he ought to be dealt with and promptly excluded, all agreed; but who was to go before the church of which he was a member and prefer the charges? Certainly I thought brethren Huckins and Tryon were the men for this unpleasant duty, and urged the point accordingly. Pressing engagements were offered as excuses for their return, and the task was laid upon the writer, or the discipline must for the present go by default, and thus allow the pernicious doctrine to gain foothold by delay. My family on the Guadalupe, and the church at Gonzales, needed my assistance and presence at once; but as no one else would remain and undertake to expose this heresy, I determined to do it, and commenced preparing the way before the regular meeting of the church.

Contrary to my natural inclination, I had often, by force of circumstances, been compelled to meet Indians and Mexicans on their invasions into the country from the west, and aid in repelling force by force, and now, equally contrary to my inclination, I was compelled, by a sense of duty to the cause of my great Master, to confront, and if possible beat back, this advance guard of heretics invading the country from the east. I attended the conference, but Elder Cox entirely ignored my presence — preached one of his heterodox sermons — had the way prepared for the reception of a large number of persons, who would say they believed that Jesus Christ was the Son of God, no other testimony being required; and when these would be baptized, he would be largely in the majority, and could turn the minority out. Having been informed from a reliable source that this was his intention, charges were written out against him, and placed in the hands of a licentiate minister belonging to the same body.

An opportunity was offered at the close of the sermon for the reception of members, when I arose and entered my protest against the proceeding, on the ground that the church was out of order, in consequence of her pastor's heresy. Elder Cox stated that this was an assumption that required testimony. The brother who held the charges then rose and read them, and it was only by a very small majority vote that the charges were entertained. Had the persons then knocking at the door for admission been received and baptized, the charges would have been thrown out by a respectable majority. It is proper here to state, that the charges alluded to were first written out and preparations made for this trial, solely on the ground of heresy. Just before going into conference, however, Elder R. E. B. Baylor sent me a paper published in Alabama, in which there was an advertisement of Cox's exclusion from the church he had left, on the ground of flagrant fraud committed about the time he left for Texas. The character of the testimony printed in this advertisement made his guilt perfectly apparent. A charge based on this, in connection with this heresy, was also presented, and a long and very unpleasant debate ensued. He was a man of extraordinary ability, especially in defence of himself, and in consequence of his commanding manner and pleasant address, taking right hold of the sympathies of his hearers, it was with great difficulty that the church could be made to see his guilt under the charges. Three times during the trial he was charged with false statements, and the proof clearly brought to bear. After a full investigation, the church was called to a vote, and he was excluded. Of course, under such circumstances, the ill will and hard sayings heaped upon his accusers, by those who still sympathized with him, were very bitter. Faithfulness to Christ demanded promptness and decision, and when afterwards I called to mind the saying that is written, "If I yet pleased men, then would I not be the servant of Christ," it gave me much consolation.

The ministers of the entire association accorded with the church in her decision, and were gratified to learn of her faithfulness. Most persons supposed that with his exclusion the difficulty was all over; but I could foresee that Cox's influence, left upon the minds of those whom he had baptized, would cause trouble in the future. Here was indeed a severe stroke upon the three little churches that formed the Union Association at its organization. Cox, up to this, was the pastor in charge of each one of them. The brethren, however, rallied around the standard of truth, and through them God has wrought wonders in Texas. Step by step, during those years of trial, truth gained upon their sympathies, and the child born upon the prairies of Western Texas grew to be a woman, and the mother of vigorous daughters, now scattered north and west.

Union Association is the mother of Colorado, Trinity River, Tryon, and Little River Association; the mother of the Baptist State Convention, and the grandmother of Austin, Leon, Richland and Waco Associations. She is the mother of the Education Society, and the Sunday-school Convention of Texas. No body of Baptists in Texas had more trials in its early history, and none perhaps has been blessed with a greater degree of prosperity.

While a dark cloud was hanging over the infant churches, in consequence of the heresy of Elder Cox, a bright star arose, and rested steadily over the old town of Washington, on the Brazos. As the wise men of the east rejoiced over the precious treasure to which the star directed them, and over which it rested, so the men of Israel in Texas rejoiced over the first extensive revival. Where men scoffed in 1837, and where the most violent opposition to the gospel was put forth, was the place God delighted to honor with refreshings from his presence.

The little church organized in 1837, in Washington, ceased to exist, in consequence of the removal of the brethren; but when brother Tryon came, in 1841, it was reorganized, and under his efficient ministry the little band delighted to read the promise, "Fear not, little flock; for it is your Father's good pleasure to give you the kingdom." The faithful pastor was confined at home, in consequence of family afflictions; but Elder R. E. B. Baylor, then holding the office of judge under the republic, was attending his court at Washington during the regular time for the church-meeting, and preached. God was in his heart and among the people, and he preached with great power.

The meeting was about to close, when the brethren recognized some unusual demonstrations, and urged Judge Baylor to continue the meeting. The Spirit of God rested upon the community, and a deep interest took hold upon almost the entire population. Elder Baylor was alone, nearly all the time, as a preacher; but

God gave him strength and wisdom equal to the task imposed upon him. The meeting continued for two weeks, and a large number professed a hope in Christ. Almost every night the ordinance of baptism was performed in the Brazos, which, interpreted, means "the arm of God."

The baptismal scenes were of the most interesting and impressive character. The moon was shining on these occasions beautifully. The congregation marched in procession, singing the songs of Zion, from the place of worship to the river. The noise went abroad of the mighty displays of God's power to save, and also of the beauty and sublimity of the baptismal scenes, and from twenty-five miles people came, and were themselves baptized before they returned. Forty-two were baptized during that meeting, and some of them yet live, to bless the church with their influence and to tell of the power and willingness of God to save sinners in 1841. Among the number baptized by brother Baylor at that time was the venerable sister America Lusk, and her daughter, sister M. E. Crumpler, both members of the church at Brenham, and who have wielded such a blessed influence for the cause of Christ in Texas since.

CHAPTER XIII.

THE DECISION, UNDER DARK CLOUDS. — 1842.

WHEN the year 1841 closed, I was at home, two miles above Gonzales, on the Guadalupe, after an absence of nearly three months among the churches between Brazos and Colorado. I of course had been sorely tried, amid the difficulties alluded to in the previous chapter. A hurried trip was made to Mississippi for my children, for want of means to educate them further. This about consumed all the means I had left. The old dreamer, John Bunyan, said that "An idle brain is the devil's workshop." A little farm was at once opened on the river, and twenty-five dollars paid for a plough to break the land. An old set of blacksmith's tools was secured, a young man of good character employed, and farm and shop, by turns, employed a good portion of my time during the spring of 1842.

Preaching was kept up regularly at Gonzales, and at a school-house four miles above. During our absence one night at meeting in Gonzales, the Indians stole the last pony we had. The horse was staked about forty steps from our door. After this we all went to meeting together from my neighborhood in ox-wagons. God blessed the little church with some precious seasons.

Once I had been alone, as a Baptist preacher, between the Brazos and Colorado. Now that Tryon, Huckins, Baylor, and Garrett were occupying the field east, I was again entirely alone, as there was not a Baptist preacher west of the Colorado River to confer with me. At this time I met with Elder Carroll, a Methodist circuit rider, travelling in the valley of the Guadalupe. Reaching my house, his shoes were worn out. There were no shoes in the country to buy, and nothing scarcely but rawhide to reset. My old Tennessee shoe tools were still on hand, and a few small pieces of leather. He came to me in the midst of his distress and inquired if I could in any way relieve him. We did not agree at all on the doctrines of depravity, baptism, communion, and church polity, but just now we were agreed on the old Tennessee doctrine, that a boy was not fit to marry til he could stock a plough and mend a shoe. His shoes were mended, and after a pleasant interview he went on his way, asking a kind remembrance of the Baptist preacher at the throne of grace, for divine protection in the midst of the dangers that hung upon his path.

While at the school-house four miles above Gonzales, at a night appointment, a scene occurred worthy of record. Some were standing guard, and others, in the rear of the congregation, sat with their guns across their knees.

I preached with unusual liberty; the attention was undivided; many earnest prayers were offered for our protection in the midst of difficulties and dangers, and some praised God aloud. The congregation was dismissed, and before leaving the place a gun was fired a few hundred yards away; the shrill Indian whistle was heard, and the people warned to proceed with caution to their homes. As the way home for all the congregation was the same for some distance, my ox wagon, carrying my own and two other families, took the lead. The others, travelling much the same way, fell into line, and we moved off calmly, with no confusion manifest. A proposition was made tht we should sing one of the songs of Zion, to drive the gloom away. Soon the echo was heard along the valley of the Guadalupe, and no doubt in hearing of the red warrior, of that old song so full of faith and heaven: —

> "On Jordan's stormy banks I stand,
>   And cast a wishful eye
> To Canaan's fair and happy land,
>   Where my possessions lie.
> Chorus. — "Oh, sacred hope; oh, blissful hope,
>   By inspiration given,
> The hope, when days and years are passed,
>   We all shall meet in heaven."

I then thought, and yet think, that amid the solmnities of that hour I heard the sweetest music to my soul that ever fell upon my ear. It was a lovely moonlight night, and a consciousness was realized that God would protect this company of worshippers to their homes.

The next morning we assembled, after news was received that a man was killed. About two hundred yards from the little school-house where we worshipped the evening before, we found the body of Dr. Witter, an eminent physician, blood-stained and terribly mangled. We buried the remains as decently as our facilities would permit. Here a little mound was raised over the body of a learned infidel, who refused to go to meeting, though it was so close to his house, and beside that grave stood four interesting children; the eldest about ten years of age, and scarcely done weeping over the loss of the mother, whom we had buried a short time before.

Just at this time a traveller from the east brought me a letter from the Colorado valley, bearing date February, 1842: —

"Dear Brother Morrell: Our conference meeting comes on at Plum Grove next Saturday. We are in trouble. The anti-missionaries have been among us, sowing the seeds of discord. We are on the eve of a rent in the body. Come and help us. You may effect a reconciliation. Come if possible; and may the Lord come with you.

"Wm. Scallorn."

The contents of this letter gave me pain, and I felt anxious to go and lend assistance; butit was now late Friday evening, my last horse was stolen, and my oxen could not carry me fifty miles by the next day in time to meet the conference at eleven o'clock.

This was once, I remarked to a young man from the neighborhood, who was waiting for work to be finished in the shop, that I could not go. He kindly tendered his Indian fighting pony for the trip, expressing a willingness to walk home two miles. In about an hour by sun, with my gun across the saddle in front, I left for a lonely night-ride. Should I fail to settle this difficulty, and the church should be rent asunder, where I had first administered the ordinance of baptism, the temptation to return to Mississippi, under which I was then laboring, would be increased. Only a short time previous to this a letter had been received, as follows: —

City of Austin, Texas, January 10th, 1842.

"Dear Brother Morrell: I have been inquiring after you all the way out from Mississippi; but as you are at Gonzales, and as there is so much danger in that direction, I cannot go down. Besides, my time is out, and urgent business calls me home. Information has reached me, that in the midst of this revolution you have lost everything you brought to the country. If you will return and labor for us, I will, if necessary, send you a check to defray your expenses back, and give you a better tract of land than you sold when you left us. We need you very much in Mississippi. Return if you possibly can.

"Truly, your brother,

"James Greer."

This letter had given me much trouble, and was yet unanswered. About twelve o'clock at night I stopped and rested one hour, and just at daylight reached brother Scallorn's, fifty miles from home. After a cup of coffee, I slept till nine o'clock, took breakfast, and met the congregation at eleven.

The church conference was well attended, and after a sermon and the usual preliminaries, the question of missions came up. A studious effort was made by certain brethren who had come in among them to make this a question of fellowship. After a lengthy conference, with a free and frank exchange of views, the church agreed to bear and forbear, allowing every brother to give to or withhold from the mission cause, as he saw proper. A sermon was preached on Saturday night, and after services Sunday morning we commemorated the death and sufferings of Christ. Here, instead of differences, the brethren manifested the warmest fellowship. Another sermon was preached at four o'clock, and late in the evening I left for another ride of fifty miles by night. My presence was demanded at home early Monday morning, and as it was very unsafe to ride over that part of the country in daylight, the trip must be made at night.

Rejoicing with my family next morning at nine o-clock over the settlement of the difficulty, we discussed the propriety of returning to the Colorado, in answer to a call from the brethern to return and occupy my former field. We had exchanged locations so often during our short stay in Texas, that we decided to remain, if it was the will of God, on the Guadalupe.

Two Cumberland Presbyterian missionaries came in a few days after this, and preached in Gonzales several days. Nearly all Christians of the town appeared revived, and notwithstanding the dangers daily surrounding us, the whole community seemed pervaded with religious influence. These preachers were brothers, bearing the names of John and Phineas Foster.

About this time the following letter was written: —

"Gonzales, February, 1842.

"Dear Brother James Greer, Holly Springs, Miss.: — Your letter of the 10th inst. is before me. Prayerfully I have considered its contents for many consecutive days. You are correctly informed with reference to my finances. It is true that very little has been done by way of organizing churches in the republic, during these seven years of my sojourn in the wilderness. Considering the difficulties under which we labor, we have much to encourage us, and we ought not to grow weary, but 'let patience have her perfect work.' My faith is strong in the final success of the republic, and the triumph of truth over error and superstition, that have long held sway over three beautiful valleys and far-reaching prairies. My first impression after reading your kind letter was to return; but after mature and prayerful deliberation, my mind is made up, that duty requires me to rise or fall with Texas and the cause of my Master in her territory. With feeling of profound gratitude for your very liberal propositions, I must decline accepting, believing it is the mind of the Lord.

"Yours in Christ,

"Z. N. Morrell."

CHAPTER XIV.

MEXICAN INVASION. — 1842.

PRESIDENT Lamar's administration closed in November, 1841, and after a most exciting canvass Sam. Houston was the second time chosen as president of the Republic, by an overwhelming majority. Colonel Edward Burleson was elected vice-president. 1842 was upon us, with the "Hero of San Jacinto" to lead us through the dark days of poverty and war that awaited us. The war-dogs of Santa Anna were howling furiously in Mexico, reporting themselves in full outfit, and determined to quell the rebellion in Texas and plant the Mexican flag on the Sabine. Scenes of blood must yet be passed over before we could in peace worship beneath our vine and fig-tree.

General Houston's policy was much opposed to that adopted under the former administration. He condemned the Santa Fe expedition on the grounds that it was impossible for Texas to hold such a large territory against so many enemies; and that the recklessness of such expeditions must inevitably injure Texas in her efforts to secure sympathy and aid from other governments. He opposed any effort to exterminate the Indians, by following them into their territory, as utterly fruitless, and favored the establishment of trading-posts along our frontier. He insisted on a retrenchment of expenditures, — deferring the payment of the debt to some remote period; and the issuing of exchequer bills, with reduction of taxes. These things inspired the people with fresh confidence, and gained for the republic respect among some of the nations of the Old World.

General Houston was personally opposed to the policy of annexation to the United States; but he foresaw that this tide of immigration from the old States, which he had all the time encouraged, would by an overwhelming majority favor it; and, instead of throwing obstructions in the way, favored annexation himself. I can speak advisedly here in consequence of an interview about this time with President Houston on a boat running between Houston and Galveston. On that trip he told me that in consequence of the rejection of the hand of Texas, tendered by the vote of 1837, and in consequence of the unmerited contempt afterwards shown, through the influence of chicken-hearted politicians east, he intended "to turn coquette for a while, and court England and France, right before the eyes of the old lover, the United States;'' and that, under the influence of a little jealousy, he thought more liberal arrangements could be made at the marriage feast. His engine of ingenuity was soon fired up, and in a subsequent conversation with him he informed me that as soon as the smoke of jealousy began to rise, the dastardly chicken-hearted politicians became more pliant.

On Saturday before the first Sunday in March, the little church at Gonzales met in conference and offered an opportunity for the reception of members. Two letters were received, and my son James presented himself, aged seventeen, as a candidate for baptism. This was the first application for baptism on the Guadalupe, and there was a spirit of rejoicing manifest. Ten o'clock Sunday morning was appointed as the hour to administer the ordinance. Many of our appointments in life, under the general providence of God, are disappointed. It was so in this instance.

On the fifth day of March, 1842 a Mexican force, supposed to be about a thousand strong, approached San Antonio, and demanded a surrender. The Texan force evacuated the place, and retired up the valley of the Guadalupe, sending expressmen ahead to notify the citizens.

The messenger reached us late Saturday evening, and, after a little consultation, it was decided that families, flocks and herds must start east, early Monday morning. Everything of course was thrown into confusion. Sunday morning's sun arose, and instead of shining upon our people on their way to the baptism, furnished them light by which to make their preparation to retreat before the invading Mexicans. My little blacksmith shop was very soon surrounded with wagons, needing repairs for the journey. More wagon wheels were repaired on that Sunday than ever I witnessed at one little shop on any before or since. Wagons were loaded on Sunday night, and Monday morning a boy from every family that had one was detailed to go out on horseback and drive in all the stock of every description for miles around. The bleating ad lowing of the herd reminded us of the roving shepherd patriarchs. By one o'clock Monday everything was in motion for the Colorado valley.

What provisions we could haul were brought with us, but these were soon consumed; and as the state of the country would not allow us to move back, some returned in a short time to work out the crops, and others engaged in such employment as they could find for a living. My money was all gone, and receiving a proposition to teach, I undertook a small school, with pledges to pay tuition in provisions and stock. Money was outof the question. Preaching was kept up regularly at Plum Grove.

While engaged in this school, a letter was sent from a church up near Austin, constituted in the house of John Hornsby, requesting me to go up and assist in the ordination of brother Richard Ellis, who had formerly lived with me in Washington County. Having no horse to ride,and depending upon my little school for a living, I wrote to the church, saying, that if by order of the church brother Ellis should come down, we would, in the presence of Plum Grove church, hear him preach, and hear a statement of his call to the ministry, and if we considered him qualified would call a presbytery and ordain him. We believed this to be scriptural, and in keeping with usages of the Waldenses and Albigenses in the days of the Inquisition. When all the preachers were sacrificed in one district they selected another, and sent him a great ways with letters applying for ordination. In no emergency would they assume the office of the Christian ministry without the observance of these scriptural ceremonies. Brother Ellis was sent down and ordained at Plum Grove, by a presbytery composed of R. E. B. Baylor and Z. N. Morrell. During the same year a similar request came from the same church to go up and ordain brother N. T. Byars. Arrangements on this occasion were made to visit the church, and brother Byars was duly set apart to his work. The name of the brother who aided me in this ordination has escaped my memory. He has long since gone to his reward. Brother Ellis preached with great power a number of years, and passed to his reward while in the prime of life. Brother Byars labored actively as a frontier preacher, with great success, for a long series of years, and was recently engaged in an active mission work on the Mississippi River.

The five months' session was taught without any loss of time from the school-room, and about the first of September, as it was not considered safe to move our families back to Gonzales, I took my wagon, attended by my son, and went to gather our corn on the Guadalupe. The corn was gathered, and just as we were starting back with a load to Colorado, Colonel Matthew Caldwell rode up with an express from San Antonio, as follows: —

"*Colonel:* — General Woll has arrived at San Antonio with thirteen hundred men. The court, — judge, jury, lawyers, — and many citizens in attendance, are prisoners in the hands of the Mexicans, I made my escape, and came round under the mountains to Seguin.

"John W. Smith."

This gentleman was well known and could be relied on. The dispatch was received on Monday morning. Colonel Caldwell said, "Something must be done quick, and you must go with me." My excuses were rendered, — I was in very feeble health, was a cripple, was riding a wild, untrained, borrowed horse, and was badly needed at home. He urged me to accompany him, stating that I could be of great service to him in controlling the young men who would be with him. My patriotism was appealed to, and remembering the sentiment contained in the letter to brother Greer, that I expected "to rise or fall with Texas," my consent was given to go on another perilous expedition.

My son started on alone with his load of corn, to the Colorado, fifty miles. Although an Indian country was between him and home, I did not apprehend danger, as men in companies would soon be on the road from the east towards the scene of action.

CHAPTER XV.

WAR. — 1842.

DURING the spring and summer of 1842, a great interest was felt throughout the Republic for the annexation of Texas to the United States, and a plea was urged that the war with Mexico was about at an end. The Mexican authorities, of course, threw every obstacle in the way of this union that was in their power, and learning that this plea was made, sent out the expedition alluded to under General Woll. Their expressed intention was to march through the territory; but their real intention was to make a raid, and thus delay, and if possible thwart annexation, hoping in the end to induce Texas to submit to Mexican Rule. On the eleventh of September, 1842, a Mexican force, under General Woll, about thirteen hundred strong, captured the city of San Antonio, making hostile demonstrations towards other points farther east.

We gathered what ammunition we could at Gonzales, and left for Seguin, with instruction that recruits coming from the east should follow our trail. At Seguin I obtained ten ears of corn, had it parched and ground, and mixed with it two pounds of sugar. This we called cold flour.

Recruits were coming in all night, and on Tuesday morning we marched on within twenty miles of San Antonio. Colonel Caldwell was in command, by common consent. A call was made for ten of the best horses and lightest riders, to go and meet Jack Hays that night on the Salado. He had notified us, by express, that he was there watching the enemy, and needed reinforcements.

The number called for was soon obtained, — the writer among them, on his fine, untrained, borrowed horse. A charge, with some instructions, was given us, and a short while before day we arrived at the spot where we were ordered to go. A keen whistle was given, and readily responded to by Hays. Wednesday morning came and found us thirteen strong, with nothing but cold flour to eat, and a limited supply of that. Our ration consisted of a spoonful for each, mixed with water. A detail was made to stay at camp, another to go down on the east side of San Antonio, and another under Jack Hays to head the San Antonio River, and go entirely round in the rear of the city, to ascertain if any reinforcements were coming in from Mexico. Hays was discovered during the day and driven back, making no discovery himself as to reinforcements. Thursday morning came, and with only a spoonful of cold flour for each, another effort was made to get the number and intention of the enemy. Caldwell still remained at his camp twenty miles east of the city, expecting the Mexicans to march on Gonzales. Hays was repulsed, as on the day before, and failed to get in the enemy's rear. The writer and part of the company went down the Salado, and discovered what we supposed to be the trail of two or three hundred cavalry, going in the direction of Gonzales. On our return we met Hays with his company, driving in some horses. Very soon, about forty Mexicans made their appearance in pursuit. We retreated until they were drawn from the timber, when, under the order of our gallant leader, we wheeled, and forty Mexicans failed to stand the charge of thirteen Texans. No damage, that we know of, was done to either party.

Friday morning, a mutiny rose in our little camp, in consequence of the condition of our commissary department. Plenty of deer and turkeys were in sight all the time, and we were all hunters; but our leader thought it best to fire no guns, and keep our position concealed from the enemy. From Monday till Friday, on a little cold flour, measured out by the spoonful, made us feel very lean; and now that the flour was all out, our men began to swear vengeance on the game at all hazards. Captain Hays insisted that I should make them a speech. I remembered the old saying, "Never try to influence a man against his inclination when he is hungry," but as my captain insisted, and as I was under orders, I determined to try. To have approached these men with a long face, and taxed their patience with a long speech on patriotism, would have been sheer nonsense. So I mounted my horse and rode out in front, with as cheerful a face as I could command, and spoke as follows: —

"Boys, when I left Colonel Caldwell's camp, I felt like I was forty years old. When I had starved one day, I felt like I was thirty-five. After that, on two spoonfuls a day, I felt like I was twenty-five; and this morning, when our cold flour and coffee are both out, I feel like I was only twenty-one years old, and ready for action. Our situation this morning is critical, — the Mexicans, we fear, have gone toward Gonzales; secrecy surely is the best policy; and we ought to report the situation, if possible, to Colonel Caldwell to-night."

An agreement was soon entered into, that we get information, report that evening, and get some game for supper.

In a few minutes we were off, and soon met Henry McCullouch with thirteen men, swelling our number to twenty-seven. Here we learned that Caldwell had discovered the enemy's trail below, and that the Mexican cavalary had retreated back to the city. The families on the Guadalupe were safe for the evening. Here was fresh beef hanging to the saddles of McCulloch's party. The company was organized on the spot, with Jack Hays captain, and Henry McCulloch lieutenant, and the young captain, with his first command, led us to the nearest water. We refreshed ourselves with this delicious beef and a good night's rest. We were camped within five miles of the city.

Before day Saturday morning, Captain Hays detailed three men, and myself as the fourth, to go in sight of the city before daylight. He took three men with him, and made the third attempt to go round the city, and was successful, bringing off with him a Mexican spy as a prisoner. Lieutenant McCulloch watched both roads leading to Seguin and Gonzales. My associates and I remained secreted near the powder-house, and before the sun mounted very high into the heavens, a Mexican came out to get a yoke of oxen, feeding near by us. As soon as it was at all prudent, we captured him and his pony, within six hundred yards of the fort, and in plain view. We could see the Mexican cavalry hastily saddling their horses as we passed out of sight with our prisoner. We rode twenty miles in about two hours, and reported to Colonel Caldwell.

The poor Mexican felt confident we intended to kill him, and on arrival at camp he recognized John W. Smith, and commenced begging for his life. He was soon pacified with the assurance that he was in no danger, if he would tell us the truth. Hays and McCulloch both preceded us to Caldwell's camp, and as some anxiety was felt for our safety we were welcomed with many cheers. The two captured Mexicans told the same story. With these statements, coming from the front and rear of the city, Saturday morning, ten o'clock, revealed to Col. Caldwell and his men the strength of the enemy. General Woll crossed the Rio Grande with thirteen hundred men, and picked up afterwards three hundred "Greezers" and Indians. Our entire force, ordered into line, numbered two hundred and two men; General Woll's Mexican force was sixteen hundred.

Saturday night we were marched to the Salado, and camped near midnight within six miles of San Antonio. Here we had much the advantage in the ground, if attacked, and during the night a council of war was held. The council decided that it would not be prudent to attack the enemy in this fortifications; but if he could be decoyed out to our own chosen ground, we could tie our horses back in the timber, out of range of his guns, and from behind the natural embankment make a successful battle, although the enemy numbered eight to our one.

Sunday morning about sunrise Captain Hays and Lieutenant McCulloch were placed in charge of thirty-eight men, to approach San Antonio and lead the enemy out. Out of two hundred and two horses only thirty-eight were found, by a committee appointed to examine them, fit for the expedition. My untrained, borrowed horse and his rider was selected to go on the trip. We reached a point a half mile from the old powder-house, and about a mile from the city, between nine and ten o'clock, Sunday morning. This was about the hour that I had for so many years been accustomed to repair to the house of God, and my position in such striking contrast gave me some anxiety. Captain Hays and Lieutenant McCulloch, attended with six men, left us, with orders to be ready for any emergency. They went down close to the Alamo, and bantered the enemy for a fight; supposing that forty or fifty mounted men would be sent out, whom our captain intended to engage in battle. Contrary to this expectation, four or five hundred cavalry turned out in hot pursuit. Hays soon approached with the command, "Mount!" We moved off briskly through the timber, and as the Mexicans went round an open way, we were about half a mile ahead when we reached the prairie. They had about fifty American horses, in fine condition, captured from the citizens and members of the court, and our horses were considerably worn with the labor of the past seven days. During the first four miles we kept out of their reach without much difficulty. Two miles lay stretched between us and our camp, and soon Lieutenant McCulloch, in charge of the rear guard, pressed close on our heels. Hats, blankets, and overcoats were scattered along our track. No time then to pick anything up. The race was an earnest one; the Mexicans, toward the last, began to fire at our rear guard, doing no damage. We reached the camp, and, when formed into line, every man was present, unhurt.

The calvary that had pursued us passed round to our rear on the prairie. About a half hour intervened, during which time we refreshed ourselves and horses with water. Captain Jack Hays, our intrepid leader, five feet ten inches high, weighing one hundred and sixty pounds, his black eyes flashing decision of character, from beneath at full forehead, and crowned with beautiful jet black hair, was soon mounted on his dark by war-horse and on the warpath. Under our chosen leader, we sallied out and skirmished with the enemy at long range, killing a number of Mexicans, and getting two of our men severely wounded. In a short time they retired, and we fell back to the main command.

Between two and three o'clock in the evening, General Woll appeared with all his infantry, cavalry and artillery spread out on the prairie in our rear, and between us and our homes. As we stood in line under the brow of the hill, the brave Caldwell informed us that he could never surrender to General Woll; that he had just returned from the Santa Fe expedition, and that it would be certain death to be taken in arms the second time. He urged us to make up our minds to fight it out, and even if it required a hand-to-hand combat, the white flag would not be raised. Closing this earnest address, he invited me to make a speech to the men. As well as my memory serves me I spoke as follows: —

"Gentlemen, — We are now going into battle against fearful odds, — eight to one, — and with artillery all on the enemy's side. The artillery can't harm us under this bank. We have nothing to fear as long as we can prevent them from coming to a hand-to-hand fight. Keep cool; let us not shoot as they advance on us till we can see the whites of their eyes; and be sure to shoot every man that has an officer's hat or sword. This will prevent them from coming into close quarters. Let us shoot low, and my impression before God is, that we shall win this fight."

Just as this time the cannon fired, and the grape shot struck the tops of the trees. The Mexicans now advanced upon us, under a splendid puff of music, the ornaments, guns, spears and swords glistening in plain view. Captain Hays' attention, as they drew near, was directed to the fact that they were intending to flank us above, and pour a raking fire down our line. Accordingly, ten men, with doublebarrel shot-guns, were detached, and stationed above to prevent it. Some of the Mexican infantry were within thirty feet of us before a gun was fired. At the first fire the whole of them fell to the ground. My first impression was that they were all killed. Soon, however, all that were able rose to their feet, but showed no disposition to advance further upon our line. Not a sword nor officer's hat made its appearance after we had been fighting five minutes. The ground on which we stood was of such a character that we could step back two or three paces and stand straight up to load our guns. The battle lasted but a little while. General Woll was at his cannon on the top of the hill, looking on; his artillery was of no use, being right in the rear of his infantry, and our men sheltered by the embankment. He could see his men falling while the Texans were entirely out of sight. The horn sounded a retreat, and the Mexicans ran away in great confusion. It was with great difficulty that the Texans were prevented from pursuing.

As the firing ceased along our line, the roar of artillery and rifles was heard in the rear of the Mexican army. We understood at once that the engagement was with reinforcements, making their way to relieve us. By the time we were up and in order to go to their assistance the firing ceased, and we knew that the Mexicans were successful.

Captain Dawson from Lagrange, on the reception of Colonel Caldwell's dispatch raised a company of fifty-two men, including himself, and came up in time to hear our guns in the fight just described. The Mexicans, being between us, discovered him on the open field and surrounded him. He rallied his men in a grove of mesquit bushes, and fought with such desperation that the Mexicans withdrew from the range of his guns and turned the artillery upon him. As there was no chance to escape, and no chance to do the enemy any damage, under the murderous fire of the cannon, he raised a white flag. The men threw down their guns, and for a while the Mexicans disregarded the surrender, and continued to send the missiles of death. Captain Dawson was cut down with the flag in his hand. When the firing had ceased, thirty-five Texans out of fifty-two lay dead on the field; fifteen were spared, and held as prisoners; two made their escape. My eldest son was one of the prisoners. This little body of men punished the

Mexicans severely, during the engagement with small arms, before the artillery was turned upon them.

General Woll reassembled his forces about one hour by sun, and standing on his cannon where it was first planted, in plain view and in our hearing, made a glowing speech to his men. The huzzas from the Mexican army were mournful in our ears. We believed then, what we afterwards knew to be true, that our friends and relatives from the Colorado were the sufferers. We could not reach him with our guns, and it would not do to expose ourselves on the prairie. The Mexicans moved off towards San Antonio about sunset, and spent the night carrying in and burying their dead in the city. A large number was killed, the exact estimate it was impossible for us to make. Caldwell lost only one man killed; no prisoners; three wounded.

The night was passed upon the battleground, — dark, anxious night to me. I learned that my son, A. H. Morrell, was in the company defeated the evening before in our hearing. Was he dead? Was he a prisoner in the hands of our cruel oppressors? were questions that revolved through my mind all night long. Three men volunteered to go with me to the "Mesquit battle-ground," and at daylight we were in our saddles. My colonel and captain cautioned me to be careful, as the enemy would certainly keep out spies; but the time for caution and fear with me had about passed. At sunrise we were on the fatal spot, examining carefully for the lost son, while two of my colleagues stood guard. Thirty-five dead bodies of friends lay scattered and terribly mangled among the little cluster of bushes on the broad prairie. I recognized the body of nearly every one. Here were twelve men, heads of families, their wives widows, and their children orphans; and here, too, lay dead the bodies of promising sons of my neighbors. The body of my son could not be found. The place was so horrible that two of the men with me rode away. One remained on guard while I continued my examination. A number of bodies were turned over before I could recognize them. One or two of my neighbors' sons were so badly mangled that I could not recognize them at all. Supposing that one of these might be my son, I examined their feet for a scar that he had carried from childhood. By this time I was satisfied that he had either escaped or was among the prisoners. I then drew a pencil from my pocket, and took down the names of the dead, so that I might make a correct report to the bereaved.

The unfortunate man of Caldwell's command who was killed on Sunday was buried with the honors of war on Monday. His grave was dug with bowie-knives. During the fight, some Indians who came out with the Mexican army approached his horse, tied carelessly some distance from the horses of our command, and he left his post, against the order of his captain, and attempted to save his horse. He killed three of the Indians in the combat, and finally they killed him, and carried off the horse. This all occurred in plain view; but we were forbidden to go to his relief, as he had disobeyed orders.

Tuesday morning our little company of two hundred and two had increased to five hundred. A messenger from San Antonio announced that the Mexicans had left for the west that morning, carrying the prisoners with them. The question of burying our dead, who fell under Captain Dawson and with him, came up. We had neither axe nor hoe, and finally decided to pursue the retreating enemy, regain if possible the prisoners, and at some future day gather up the bones of our dead and bury them at Lagrange. This was afterwards done, and a monument placed over them.

Orders were given at once, and preparations made to pursue the retreating enemy. The honorable Judge Hemphill accompanied me to San Antonio, to look after news from my boy, while the main army crossed the river above, and went directly in pursuit of General Woll. We visited Mrs. Jakes and the English minister's wife, Mrs. Elliot, who had a list of the prisoners' names. My son, A. H. Morrell, was certainly among them. The Mexicans had robbed them of their clothing; my son, on his arrival in San Antonio, was in his shirt-sleeves. Mrs. Elliot took a green blanket-coat off of her son, and put it on mine. This coat, he afterwards said, was the means of saving his life. My son was reported by these ladies as carrying a wound from a lance in the engagement, though not serious. After he surrendered, two Mexicans pursued him with lances. As a lance was hurled at him, he dodged it, but as it passed it glanced his left arm, near the shoulder. He only saved his life by running in this defenceless condition round the horse of the Mexican colonel, Corasco, who drew his sword and drove his pursuers from him. We procured some provisions, what powder and lead our horses could carry with safety, and over took Colonel Caldwell, camped on the Madina, some twenty-five miles from the city.

Wednesday evening, September 21, the Texan army came up with General Woll's rear-guard at the Hondu. Here a trap was laid for us. Our spies were out, right and left of the road and in advance. The rear-guard of the Mexicans was in the bottom, in a bend of the creek, and concealed. The Mexican general had offered five hundred dollars for the head of Captain Hays, and just at this time he came very near losing it. With all his vigilance he was here surprised. Luckey, a noble man, was riding by his side, on a finer-looking horse than Hays, and was shot through the right breast, the ball coming out at the point of the right shoulder. His horse ran about one hundred yards, and left his wounded rider on the ground. Captain Hays requested me to go to his relief, as he feared he was killed. Like all other severely wounded men, he at once cried for water. Judge Hemphill fortunately had some at hand, and it was given him. Luckey did not die, as we feared he would, but survived this severe wound, and was afterwards a member of the Senate of the Republic.

By this time Colonel Caldwell had formed a line of battle, and as no one would volunteer to take care of Luckey, a man was detailed. A fight was at hand, and every man was aware of it, and ready for action. A call was made for volunteers, to increase Captain Hays' company to one hundred men, for the purpose of charging the cannon planted on the road four hundred yards in front. General Mayfield made a speech for volunteers, but not a man responded. He was a man of ability, and could make a good speech, but his was the "voice of a stranger." Colonel Caldwell knew his men, and knew that speeches were not so much in demand as example. He knew that my son was a prisoner in the enemy's lines before us, and that Z. N. Morrell's soul was fired as it never had been before. My colonel requested me to ride down the lines, and encourage the men to come out. I galloped to the lower end of the line, with my old fur cap in my hand, recognizing and being recognized by almost every man I passed. The feelings of that moment need no description. They could not be described. My dear boy was upon the hill, perhaps in irons, and unless that cannon was charged and silenced, the sad news must be borne to his mother, that *our* Allen was in chains, in a Mexican dungeon. Halting in an eligible position, so as to be seen and heard by almost the entire command, I waved my fur cap, and spoke about as follows: —

"Boys, — You have come out here from one to two hundred miles from home, to hunt the elephant. He has been running from you for two days. We have got him in close quarters, just up on that hill. We want forty men to join Hays' company. With one hundred men, we can successfully charge and capture the cannon, and turn the grape shot the other way. The old fellow can't hurl his missiles of death at us more than two or three times before we will stop his breath. Besides, the prisoners —" and as I stood pointing my finger voices were heard along the lines, "Come, boys, we will go with him." More than the number called for were soon in line and ready for the charge.

We had the greatest confidence in our chosen leaders, Hays and Henry McCulloch. Both were cool, daring men; neither of them I suppose was over twenty-five years of age. Captain Hays was, by profession, a surveyor. His great courage and deliberation were first discovered while engaged in his profession. Six men, with Hays as their leader, were out surveying a short time previous, when a body of Indians attacked them. The determined young surveyor, with compass in one hand and gun in the other, continued to take his observations, and at the same time fire upon the Indians every time they drew near. The work was not ceased till the line was finished. This incident had much to do in securing his first position as captain.

Henry McCulloch had always been among the foremost to meet the enemy on former occasions, as cool and daring as our captain, and greatly endeared to the men by his uniform kindness and social qualities. He was not easily roused, but when stirred was powerfully wrought upon, and had not the fear of mortal man before his eyes.

Under this leadership we faced that cannon, while receiving orders when to discharge our guns, and at what point to countermarch, eagerly waiting the forward command. At length the shrill, clear voice of our captain sounded down the line, — "Charge!"

Away went the company up a gradual ascent in quick time. In a moment the cannon roared, but according to Mexican custom overshot us. The Texan yell followed the cannon's thunder, and so excited the Mexican infantry, placed in position to pour a fire down our lines, that they overshot us; and by the time the artillery hurled its canister the second time, shot-guns and pistols were freely used by the Texans. Every man at the cannon was killed, as the company passed it. How many of the enemy were killed and wounded besides these, we had no means of ascertaining. Had the Mexicans charged us along the road we followed, and given us the position they occupied, but very few would have returned to tell the story; but, strange to say, they were so frightened that they entirely overshot us, killing only one horse, and wounding one man. My friend Arch Gibson, one of my nearest neighbors on the Guadalupe, who was riding on my right, lost his right cheek-bone. To prevent him from falling and being trampled to death, I threw my right arm round him, seizing the rein of his bridle with my right and guiding his horse and mine at the same time, bore him safely to the rear, in a speechless condition. His first cry was for water, which was furnished as quickly as possible. He recovered from his wound, and was afterwards doubly my friend.

The night was now coming on, and the firing ceased. Most of the men were anxious to charge the lines, and reach the prisoners at all hazards. Ben. McCulloch, who had acted as captain in other engagements, a gallant and safe leader, but who from some cause did not get into our organization in time to be placed in command, after an examination of the enemy's position, advised that the attack be postponed till morning. A sad night to me it was. Will the prisoners be retaken? Or shall they wear out a miserable existence, amidst the rattling of chains? God forbid that any minister of the blessed Jesus should ever again be driven to such desperation as I then felt! I was prepared for almost anything, as the morning will show.

During the night General Woll moved off in our hearing, and in the morning at sunrise his drum sounded in my ears about six miles on the prairie beyond. The men were called up early in the morning, knowing that a council of war had been held, and that Caldwell was advised to lead his command in pursuit of the enemy. Feeling anxious to overtake the enemy early in the day, lest the coming night might interfere with the capture, as on the evening before, I did all I could to assist both Hays and Caldwell to get the men ready.

General Mayfield, who had made an unsuccessful speech the evening before, called the men around him and commenced a harangue. He told them that we were in an enemy's country, that the Mexicans more than doubled our number, and that General Woll was hourly expecting a large reinforcement. In the midst of these dangers he doubted exceedingly the wisdom of the pursuit. His design evidently was to kill time and discourage the expedition, in same speech. My indignation now passed all bounds, and it would not be too much to say that I was abolutely *furious.* He had no command, and I had none; so that as private soldiers we were on equal footing. In the midst of his speech I interrupted him, saying that the time had passed for long speeches, and that I, for one, would be better pleased to hasten to the fight and recapture of the prisoner boys. I pointed to the baggage wagons and the cannon we had captured the evening before, and urged the pursuit. Seeing that the men were many of them about to waver, and being in perfect sympathy with my cause, the Honorable Judge Hemphill, and others of like spirit, wept at my side. In spite of all that Colonel Caldwell, Captain Hays, and others could do, the contest was abandoned. It required at this time the combined strength of our little army to compete with the enemy, and as Mayfield had succeeded in initimidating quite a number of the command, it became necessary to give up the pursuit. General Woll reported to his government that he lost on his campaign six hundred men; so that at the time we allowed him to escape he did not have more than eight hundred men. Five hundred such Texans as ours could easily have killed and captured the whole army. This was certainly one of the most disgraceful affairs that ever occurred in Texas, and this I suppose is the reasdon why so little has been said of it in the public prints of the country. The poor boys were carried to prison and chains, and we saw not their faces again for two years.

We now dispersed in small companies and took up the line of march for our respective homes. Gladly would I have hid myself from my neighbors, if duty would have permitted, rather than rehearse the sad story relative to their dead, and the manner in which they were necessarily left on the "mesquit" battle ground to be devoured by the crow and the wolf.

Heaven I hope has forgiven me for the animosity I felt towards the man that made the long speech. Twice-afterwards he approached me in a friendly manner. The first time was on the return home. I replied to him by laying both my hands on my gun, forbidding him to speak another word. This may have been wrong, but I did it.+ The second time he approached me was on the streets of Brenham, Washington County, Texas, years afterwards. God had caused my poor heart, in the mean time, to bow beneath the greatest affliction in life, and I tendered General Mayfield my hand, and endeavored to *look* forgiviness, — I did not feel like talking. My wife was in the grave, hastened there prematurely, as I believed,

+I relate this last incident, and some others, partly in self-defence, against the charges implied in some pleasant anecdotes told among my friends, in which there are exaggerations, and many things derogatory to ministerial character. Facts are given in all these cases in accordance with the most rigid taxation of my memory.

by the grief of two years, in consequence of the chains her eldest child wore in a foreign land. When he questioned me as to my feelings towards him, faithfulness required me to say, that there were some wounds made in life that could not with safety be probed, even when they were old; and that *this* was one of them.

CHAPTER XVI.

AFFLICTION. — 1843.

SELDOM in the history of human affairs does a people pass through a more gloomy period than we experienced in the fall of 1842. Widows among us wept and refused to be comforted; mothers mourned in consequence of the imprisoned and dead, and there was anxiety lest the widow and orphan should suffer for bread. Our crops on the Guadalupe were all consumed, whether in the field or in the crib, by the passing soldiery, and to us signs appeared forboding war on a large scale. As, however, a calm follows a tempest, and as sometimes the highest joy succeeds the deepest sadness, so Texas passed the last trials, in September, of an invasion of her territory by any large force of her enemies. Occasionally afterwards there were conflicts between small parties, but this was the beginning of better days for the Republic. The dim rainbow of promised peace very shortly spanned the heavens, and as the sound of war gradually died away, the gospel trumpet sent her silvery notes across our plains. Hence, in the future, we will be permitted to record less of war, and more of religion.

The monthly meeting was just at hand, and delegates were appointed to the Union Association, to meet with the church at Washington, on the Brazos River. The church first organized there had disbanded, but another organization had been formed, under the ministry of brother William Tryon. Arriving at brother Farquahar's, close by the place, we were informed that the association would not meet. The people and brethren at Washington, in consequence of the great poverty of the country, had decided that they could not sustain it. I felt considerably provoked, and gave vent to a little of my displeasure. Having just returned from the western campaign, where we had together starved three days at a time in defence of the country, patience under the circumstances was more than the brethren, in their charity, expected of me. My language, as well as I remember, was that "A set of Baptists that could not live on beef alone, in times like these, through the short session of an association, were not worth shucks." Brethren now would get offended with such talk; but we did not mind it much then. If a brother got a little mad about anything we did, we just let him blow out, knowing that he would feel better when he got in good humor. We held a little conference with brethren Tryon, Baylor, and others, and called a meeting of the association at Mount Gilead, near the present locality of Brenham, Washington County.

The association met at the place appointed on Saturday, the twenty-sixth of November, 1842. The business transacted and the acquaintances made at this meeting greatly encouraged us, notwithstanding the distressed condition of the country at large.

Here I met for the first time Elder Hosea Garrett, just from South Carolina, his native State, whose face has appeared at nearly every session of the Union Association since, and whose counsel has been as wise, upon the whole, and received at the hands of the brethren as much consideration, in all our deliberative bodies, as that of any other man that ever came to Texas as a Baptist preacher. At first, he was considered more modest and retiring in his manners than was best in such an era of our history; but with a warm and generous impulse he steadily maintained the dignity of the Christian character, and won his way to confidence and position. Although he did not enjoy the privileges of an early education, he possessed the rare quality of good common sense. He has ever been the true and steady friend of our literary institutions, as all his past record shows; and although as a man, and as a preacher, he has at no period of our history appeared as a blazing comet, yet as a steadily shining star he has all the time faithfully reflected his borrowed light.

Seven years with me had passed, — years of war, attended with frequent changes of plans and locations, — and, weary of frontier life, my mind led me to seek repose. The welfare of my helpmate required it. Her spirit was crushed by the previous loss of our elder daughter; and now that the elder son was in chains, and in the hands of a semi-savage people, her health was rapidly declining, and I could under these circumstances no longer join my countrymen in absences from home. My way was by no means clear. The church I first organized at Washington failed, and now the frontier church at Gonzales were scattered; my farming and financial operations all had failed, and in the midst of my distresses, like Jacob, after the loss of Joseph and Simeon, and the demand for Benjamin also, I could but cry out, "All these are against me;" and faith revealed no reason why these things should fail to "bring down my gray hairs with sorrow to the grave." We moved to the city of Houston, and after a few weeks of patient search for a field of usefulness, in connection with means of support, I located the family near the mouth of the Trinity River. Here I preached what I could among the scattered settlements, and waited the developments of the country.

My cup was full of sorrow, but the Father of mercies and goodness determined that it was best for me and for his glory to make it run over. The partner of my bosom sickened, suffered long, and died. The son she longed to see again was seen by her no more on earth. She went to sleep with the clearest hopes of heaven. Had it been God's will, I could then cheerfully have taken the three remaining children to join her and my Master on the other shore.

In my age I have nerved myself, with the blessing of God, to write the trials of mine and others, endured in 1842. Clouds of gloom hang so heavily around all the recollections of 1843, in the west, that I must leave its records to others. I can't write them.

CHAPTER XVII.

DISSENSIONS AND TROUBLE IN EASTERN TEXAS. — 1843.

AS the star of empire has ordinarily made its way west, and as it has been a custom to go from west to east in search of light and civilization, it may appear strange to the common reader why Eastern Texas made slower progress, in its moral and religious developments, than Middle and Western Texas. West of the Brazos River, previous to 1843, we find a number of churches organized, and an association of churches meeting annually, after 1840, with several earnest, consecrated preachers. It was three years after the Union Association was organized in Washington County before the Sabine Association was organized in Nacogdoches County.

Although Elder Isaac Reed came to Texas and settled near the town of Nacogdoches in 1834, more than a year before I went to the west, and notwithstanding his ability as a preacher, his zeal and personal piety, no church was organized previous to 1838, and after this, for several years, but little success crowned the work of this good man and his associates. With Elder Reed I was personally acquainted, and labored with him in the western district of Tennessee. He there served as moderator of an association; many baptisms and large success attended his ministry there. As the common enemy did not distress by invasions that part of the country where his labors were first given in Texas, there must necessarily have been some reason otherwise why the cause he maintained advanced so slowly. The real cause certainly existed in the peculiar state of the society bordering on the east. Good people, it is true, lived there; but a large part of the population was less easily impressed with the gospel than the families in the west, who were constantly being driven from one point to another by the Indians and Mexicans.

It is well known that a large tract of the country northwest of Nacogdoches was in possession of the Cherokee Indians previous to 1839, with but few white settlers among them. It is also a notorious fact that all that portion of the country bordering on the Sabine River was for a great number of years known as *neutral ground.* The boundary line between Texas and the United States had not been definitely agreed upon. A large number of refugees and desperadoes infested this neutral territory. They crossed the Sabine River to escape justice in the United States, and recrossed it if pursued, for violations of law, by the Mexican authorities. The little river, winding its way through this neutral ground, was long considered the natural savior of thieves, and robbers, and murderers. A counterfeit spirit grew luxuriantly on that soil. In 1838, when the Texas land office was opened, and certificates began to make their appearance and gradually increase in value, this band of Satan commenced the manufacture of a large amount of false certificates, which were sent west and put upon the market. The writer unfortunately purchased one of these certificates for a friend, for five hundred dollars. The society of that section was long cursed with the presence and influence of this band, which was composed of men of intelligence, and who were sworn enemies to morality and religion. Their imprints were left upon the rising youth of the country, and it was not in the power of man to prevent it.

In 1842, the scheme was exposed to the public gaze, and a contest opened between honesty and rascality, in Shelby, and the adjoining counties. Blood was spilled on several occasions, and the courts, instead of executing the laws, fled for safety, leaving society in a fearful state of anarchy and confusion. A party of citizens arose in arms to check the infringements of the lawless bands roving through the country. This party, however, was only a mob, and prosecuted their cause with so much zeal that great injustice was in some cases dealt out to innocent citizens. These were called *Regulators.* Another body of men was soon organized, to oppose the extreme measures of the Regulators, and these were called *Moderators.* These parties continued their strife for several years. It was almost impossible for any man to remain in that section of the country, without taking sides with one or the other of these parties. In 1844, General Houston sent a body of militia to the scene of action, which succeeded in influencing the parties to lay down their arms and submit to the laws. Through all these scenes of bitterness and conflict a few earnest servants of the Most High toiled on, and with some success.

Elder Isaac Reed settled nine miles north of the town of Nacogdoches, in 1834, and preached as regularly as he could in that vicinity till 1838, when the Union Baptist Church, which still lives, was organized, with seven members. Elder R. G. Green assisted in the organization. Elder Reed, the pastor of this little flock, although full of the mission spirit, was opposed to boards and missionary societies, and the church, called Union, was at first opposed to missions. It afterwards became a missionary body, and is yet.

Elder Asa Wright, who still lives in Western Texas, emigrated to Texas in 1839, and labored with the writer in the same year, on the Colorado, in that precious revival at Plum Grove, where both his sons, J. V. and Wm. T. Wright, were convicted of sin. He was a man of earnestness and power. Owing to the troubles visited upon the settlements in the Colorado valley, he moved back to the east the same year, and co-operated with Elder Isaac Reed.

Elder Lemuel Herrin moved to Texas in 1841, and settled in Harrison County. Under his ministry a few Baptists were gathered together, and the Border church, Harrison County, was organized with eight members in 1843. Brethren Herrin and Reed composed the presbytery at the organization.

In the year 1839, the old Union church, in Nacogdoches County, enjoyed a revival meeting, under the labors of its beloved pastor; when quite a number were baptized, — the first ever baptized in the east. It will be remembered that the same year the writer was baptizing in the Colorado; and while the east and west joined hands under God in this glorious work, it is a question I cannot decide, whether Isaac Reed or Z. N. Morrell baptized the first candidate in Texas. In spite of the opposition of the prince of the power of the air, that worketh in the children of disobedience; in spite of his minions, thieves, counterfeiters and desperadoes; and in spite of that reign of terror, during the career of regulators and moderators in the east, the gospel was made the power of God unto salvation, and Christ's servants rejoiced in the midst of their trouble. Other churches besides Border and Union were being organized, and in 1843 there was a necessity felt for a general organization, which resulted in the formation of the Sabine Baptist Association.

This body was organized in November, 1843, at the old Union church, in Nacogdoches County, with five churches. Messengers were present from Union and Mount Zion, Nacogdoches County; Border and Bethel, Harrison County; and Bethel, Sabine County. The ministers who took part in this organization were Isaac Reed, Lemuel Herrin, and Asa Wright. The churches composing this body were small, but among them was an aggressive element, notwithstanding the internal commotions from which they suffered. In 1846, their minutes show a membership of three hundred, with Isaac Reed as moderator, and in 1847 they numbered five hundred and twenty-seven, with William Britton as moderator. While in the association and among the churches west of the Brazos the admirers of Alexander Campbell were giving us trouble, the brethren east of the Trinity were suffering sorely in consequence of the anti-missionary element. One extreme, if pressed persistently, usually begets another, and this furnished no exception to the rule. Antinomianism, founded on predestination and election, pressing the eternal purposes of God, without the proper consideration of the means leading to the end, drove some brethren to the opposite extrme. who, under the influence of Arminianism, waged a relentless war against their "iron jacket" brethren. These opposing elements, both alike at war with truth, finally resulted in the dissolution of the Sabine Association, at its sixth or seventh session, held with Mount Olivet church, Cherokee County. The anti-missionary and free-will elements went off into small and separate organizations. The mission element rallied under the auspices of the Soda Lake Association, which we will notice at the proper time.

In 1844, a convention was called by the regular Predestinarian Baptists of the East, which met with the Antioch church, in Jasper County, on the eighth day of November. Five churches were represented in this convention, viz.: — Antioch, Louisiana; and Salem, Antioch, Harmony and Mount Olive, Texas. This convention appointed, the same day, a committee to report articles of faith and a constitution, which report was read and adopted on the morning of the ninth. The caption of the report read as follows: — "The Articles of Faith of the Louisiana and Texas Regular Predestinarian Baptist Association."

Elder Levi A. Durham was their first moderator. He was a man of great originality; thought strictly for himself on all questions of theology, and boldly preached what he believed. I have met but few men in life so well versed in the Scriptures. He was a man full of zeal in advocating his views, and during my intercourse with him, I was favorably impressed with his personal piety. In 1845, about a year after the organization just alluded to, I met him at Owensville, in Robinson County, during the session of the court, brother R. E. B. Baylor presiding as judge. We preached alternately for several nights, and in these sermons discussed fully those points of doctrine relative to which we differed. The judge and the bar manifested much interest in this discussion, giving us their regular and earnest attention. Elder Durham opposed, with all his might, all secret organizations, benevolent societies, and missionary boards, giving his special attention to Baptist organizations that granted membership upon a moneyed basis. While he thus opposed the plans upon which we proposed to send missionaries into destitute fields, in the very midst of his opposition he would occasionally manifest as earnest a missionary spirit as those who clamored loudly for boards and money. He was not opposed to spreading the gospel, but the plan upon which we proposed to do it. That the association over which he presided should oppose missionary organizations, we would naturally expect. The eleventh article of its consitution reads as follows: — "Having for years past viewed the distress that the following institutions or societies have brought upon the churches, that is to say, Missionary Effort Societies, Bible, Baptist State Conventions, Temperance, Sunday-school Unions, Tract, Ministerial, Education Societies, and, in a word, all the human combinations and societies of the day, set up in order to advance the Redeemer's kingdom, as inimical to the peace of Zion, and calculated in their nature to cause schism; we therefore declare non-fellowhip with all such."

The sixth annual meeting of this body makes a showing upon its minutes of only six churches, with a total membership of seventy-three; Elder B. Garlington, moderator. The minutes of its tenth session, held with Salem church, Tyler County, show the same number of churches, and a smaller membership; Elder R. F. Gibson, moderator. It is painful thus to witness the decline of churches, over which good and true men have the oversight. But as Christ when on earth was led by a mission spirit, and infused the same into his early followers, we should ever be impressed with the great truth that the Christian spirit is aggressive and consequently missionary.

The extreme meaures adopted by these brethren in their opposition to all mission organizations drove other brethren off to the other extreme, even into fanaticism, under the name of "Free Will Baptists."

These gave man more to do than the Bible allowed, while the others placed less upon his shoulders than it required of him. The exact date of the organization of the "Free Will Missionary Baptist Association" I cannot give, but the minutes of October, 1850, show that it met with the Ayish Bayou church, in San Augustine County, Elder G. W. Slaughter as moderator. The churches composing the association in 1850 were four, — Ayish Bayou, Bethel, Milam and Sardis. So zealously did these brethren advocate instrumentalities, that the following resolution appears in the minutes alluded to:—

"*Resolved,* This this association recommend to the prayerful consideration of all the friends of the Redeemer, that, in place of building tents out of wood on such occasions, each head of a family make a tent of cloth, and take their wagon, with forage enough to feed their horses for a few days, and enough of light diet to feed their families, and approach the door of the sanctuary, as the Israelites did the tabernacle, and take God at his word, and lay hold of his promises, and see if he will not pour you out a blessing, that will fill your heart with gladness, and make you rejoice in place of mourn when you come to press a dying pillow."

Trembling under a sense of their responsibility, and aroused by the inactivity of their predestinarian brethren in their very midst, the messengers of these four churches passed a second resolution in favor of missionary combinations and extensive operations:—

"*Resolved,* by this association, That it is our prayerful desire to see three thousand six hundred missionary boards organized in our bounds, and see flowing therefrom, as a river, the streams whereof make glad the city of God, one hundred and seventy thousand itinerant preachers, such as Paul and Silas, going forth in the name of Israel's God, conquering and to conquer, till iniquity becomes ashamed and hides its head. Then will the church of the living God come together like Solomon's temple, without the sound of a hammer."

Distressed by what they termed illiberality on the part of the Regular Predestinarian Baptists, touching church order and the atonement, they aspired to an enlarged liberality, and passed another resolution:—

"*Resolved,* That the ministers and deacons of other associations are respectfully invited to be in attendance in all our churches and meetings that are convenient, and that all Christian ministers shall receive a cordial welcome in our stands, at any time, of every denomination.''

But little progress was ever made by this organization. The leaders and followers alike possessed a zeal without knowledge, and if they have maintained an organization in later years, I have been unable to find any published statements of the fact.

The reasons now are plain to every reader why the cause of Christian missions and education in the east made such little progress, previous to 1850, among the Baptists. The faith and practice held by some of the most influential early ministers, as shown in this chapter, had much to do with it. The terrible social disorders through which they passed, previous to 1844, and the bitterness that followed for years afterwards, were obstructions of great magnitude. The conflict of extremes that hung around the early churches, and the old association, at every meeting, were impediments calculated to discourage and prevent organized effort; and in some parts of the east, till this day, these old influences are still at work, greatly hindering the prosperity of the churches. No people, in any country, free from persecutions unto death, have struggled against more formidable difficulties in the way of progress in building up the Baptist cause than these people, and yet a large number have held steadily on to old landmarks of doctrine and aggressive practice, as will be seen in the future development of their history as a denomination.

CHAPTER XVIII.

ORGANIZATION IN MIDDLE TEXAS. — 1844.

THE prospect of an early settlement of the difficulties between Texas and Mexico was very favorable in the beginning of 1844. The question of annexation to the United States was receiving the earnest attention of both governments, and emigration to the Republic was determined upon and put into execution on an extensive scale. The government, through her agents, under the pacification policy of General Houston, formed a treaty of peace with a number of the most hostile and troublesome tribes of Indians. Our finances, under this administration, were in a far better condition, and everything indicated a better state of things.

Emerging from eight years of war and reviewing the scenes through which I had passed, disappointments, removals, afflictions, in person and family, and the loss of every crop I planted, my path appeared behind me through deep waters and fiery trials. The waters had not overflowed me, and fires had not consumed me, and with a heart full of gratitude to Him who walked in the presence of Nebuchadnezzar, with Shadrach, Meshach and Abednego, in the midst of the fiery furnace, and who, walking upon the water himself, caught the hand of a sinking Peter and restored him to his place, I buckled my armor on, and, with a fixed determination to fight his battles while I lived, went forth in what I supposed to be the line of duty. With a full realization of the fact, that the sparse settlements would very soon be populous neighborhoods, in need of churches and a regular ministry, I sold out my little estate at the mouth of the Trinity, and gave myself exclusively to the work, commencing at the Providence church, Washington County.

Here I met my son, A. H. Morrell, concerning whose capture by the Mexicans under General Woll, in 1842, I have written in another chapter. For nearly two years he had been absent as a prisoner, during most of which time he was wearing chains in Perote, Mexico. In 1842, in consequence of the brutal treatment of American prisoners in Mexico, Texas requested the appointment, by the United States, of a special minister to visit Mexico. As the captives had many friends about the capitol who urged the assistance of the United States, President Tyler appointed General Waddy Thompson, who promptly repaired to Mexico; and through the influence of this minister my son and a number of others were released, early in 1844, and sent to New Orleans. On his arrival he of course rejoiced in being permitted to walk unmolested the streets of an American city; but, nevertheless, in his ragged and penniless condition, he needed friends, and found them. Of whatever sin I may have been guilty in the past, God has saved me, in the midst of my trials, from the sin of ingratitude. Gratitude lingers in my bosom still to my old friend and brother, H. C. McIntyre, of Brenham, Washington County, Texas, for the kind attentions rendered my boy, in providing comfortable quarters for him in the city, and lending pecuniary aid in his passage to Texas; also to Elder Whipple, of the Methodist Episcopal church, for pecuniary assistance on his trip from Houston to Washington County. The emotions with which I met my boy cannot be described, and only those can appreciate them who have been in like circumstances.

Before deciding which should be the field of my labor, I felt inclined to revisit the valley of the Colorado, and commune with the brethren at Plum Grove and surrounding country. Here everything was wearing a brighter prospect. Immigration was lending encouragement to the little bands of Christ's disciples, and the progress of Christianity and civilization in that lovely country was plainly visible. Elder R. E. B. Baylor, then residing at Lagrange, accompanied me to Colonel Richard Jarman's, some seventeen miles south-east, where we preached for several days and organized a church. One was received for baptism, and after an urgent solicitation on my part brother Baylor consented to administer the ordinance. He may have administered the ordinance before, but the impression on my mind is that it was the first. Greatly encouraged with the prospect, religiously, west of Brazos, my mind was impressed strongly that my labors were in demand in the county of Montgomery, which then extended from the Brazos to the Trinity River, and embraced at that time all that territory now included in the counties of Grimes, Walker, Madison and Montgomery.

I visited, as rapidly as I could, a number of the most important points in this region of country, the present locality of Anderson, Colonel Shannon's, Montgomery, Danville and Huntsville, inquiring after Baptists, and considering the facilities and difficulties relative to a general organization. The country was being rapidly settled, and large congregations met us at every point. Many of the difficulties, related in connection with organization east and west, were met with in what we call middle Texas. Although these Baptists had no discipline save the New Testament, and, in keeping with the true church in all ages of the Christian history, rejected the traditions of men, and insisted upon organization under the revealed laws of Christ, yet, coming as they did from so many different localities, and surrounded in the new country with such a diversity of interests, they were all full of notions. Times and localities for organizations, with some slight differences on doctrines, furnished the grounds of disagreement; but generally there was an earnest desire manifest to organize churches and secure the privileges of a regular ministry. Outside of Galveston and Houston there was not, at this time, a single Baptist church between the Trinity and Brazos Rivers, from their springs in the mountains to the Gulf of Mexico, that I had then, or yet have, any knowledge of, unless there was near Springfield a little antimissionary organization, and my impression is that this body was formed at a later day.

Elder James Huckins came as a missionary to Galveston in 1840, and on the thirtieth day of June, the same year, constituted the church in Galveston, with twelve members. He was appointed to labor in the cities of Galveston and Houston. On the tenth day of April, 1841, he organized the church in Houston, with nine members.

On my first trip through the country referred to, I met with Dr. R. Marsh, the old Baptist preacher, with whom I met in 1837, in Houston, and who was, with me and others, a member of the Vigilance Committee, organized the same year. He was then and had been living for two or three years in the vicinity where Danville is now located. He was over seventy years of age, and as his memory was failing, had to be reminded by the brethren of his appointments. He had been in earlier life a man of ability as a preacher, and at this time, when he could fill his appointments, preached with a considerable degree of system and power. I labored with him in that vicinity frequently, during the year, and assisted in the organization of a little church, late in the season. The date of this organization I do not remember. It was dissolved in a few years, and reorganized by brother Creath, at Danville.

The church at Huntsville is the oldest in middle Texas, and was organized by Elders Thomas Horsely and Z. N. Morrell, with eight members, on the sixteenth day of September, 1844. T. G. Birdwell and wife are the only members now living who were in the organization. But few churches in the history of Baptists have maintained their organization so long and amid so many trials as the church at Huntsville. The writer served as pastor one year after the organization, and although the church has enjoyed the pastoral care of some of the best preachers that ever came to the State, — the zealous and untiring J. W. D. Creath; the prudent and far-seeing G. W. Baines; the clear-headed and warm-hearted S. G. Obrien, — yet, at different times, it has passed through ordeals that tested severely the faithfulness of its membership. Under it all, and through it all, it lived, and yet lives, to bear testimony for Christ. My association with this church as its first pastor, and the cords of friendship and fellowship woven in after years between my heart and the hearts of many of the members of that community, have always caused my prayers to ascend for that flock.

The trials through which we passed, previous to the organization, were only second to those endured at Washington, in 1837. There was no organized effort to prevent the public ministry of the word, but a spirit of disorganization met us at the very beginning. Elder James Parker, the zealous advocate of what he called "primitive order," who was the brother of the famous Daniel Parker of "two-seed" notoriety, was preaching in the vicinity, and laboring to organize upon principles opposed to the mission work. With his principles and doctrines all dyed in waters of Antinomianism, he put forth the strength of his entire influence in opposition to organization upon correct principles. On the ground was another Baptist preacher, by the name of McClenny, who was in his views decidedly missionary, and willing to aid in organizing; but he was unfortunately, at the time, engaged in a terrible lawsuit; his enemies charging him with being accessory to murder. While it was not our business to decide upon his guilt or innocence, he was evidently not "of good report among them that are without," and with all his affability and apparent humility he was an embarrassment to us instead of assistance. Last, but not least, was Elder R. G. Green, who aided, in 1838, Elder Isaac Reed in the organization of the Old Union church, the first in Eastern Texas. I knew him in Tennessee, both as a lawyer and as a minister, when he was in good standing in a Baptist church. Trouble assailed him of a domestic character, and, giving way to evil, he became a wreck in Texas. On his arrival in Huntsville, and previous to any disorder on his part that we knew of, he was solicited to preach, and did so on several occasions. His talents commanded for him the respect of the community, for his ability was second to but few. But while in the enjoyment of this respect and confidence, the remembrance of his troubles in the old State drove him to madness. Alas! he visited the ante-chamber of the pit, sought to quench his trouble with liquid fire, and, in his struggle with John Barleycorn, was wallowed like a brute in the streets. With all his education, he walked through the town in his frenzy, and, in the presence of the people, imagined himself a steam engine, to the great amusement of the wicked, and to the great motification and discouragement of the few pious souls who panted for an organized effort against the powers of darkness.

There were about twenty persons in what is now called Walker County calling themselves Baptists; but it seemed almost impossible to get enough of them together who were willing to organize. Satan laughed at us in our efforts, and stirred up his imps in human form to tantalize us, by pointing at fashionable Baptist women in the ballroom, running the giddy round, excited by music, among some of the most abandoned characters. They probably did not know the real character of some of those who took them by the hand, in the midst of the whirl and the dance; but they ought to have known it. Troubles of vast amount have been met with in the history of many Baptist churches in Texas, in consequence of lax discipline with those who, by their actions, testify that they love the ballroom better than they love the church of Christ. The experience of the past should be warning for us in future. Who but a pioneer preacher can appreciate the embarrassments and feelings of fearful responsibility, in the midst of such pressure brought to bear by the enemy from so many directions at one time? With earnest cries a few of us approached the mercy-seat, and, realizing that man could not drive these clouds away, we waited for God.

Regularly the monthly appointments at Huntsville were filled, and although large congregations assembled on every occasion, months passed and no light appeared. Men were then guilty of collecting in the church-yard on the day set apart for worship, and instead of entering the house and giving respect to God and his servants, who labor for the public good, spent the sacred hours, to the great annoyance of the preacher and congregation, in rehearsing idle and mischievous things. If any man or youth, who reads these lines, is ever tempted to hang round the church-door during the hours of service, let me remind him that it is in bad taste, opposed to good manners, and a public nuisance. Either hide your face from the view of civilized people or take your seat quietly in the congregation, as a gentleman, till the service closes.

Men disposed to spend the time in this way had more temptations in 1844 than now. Hundreds of interesting incidents were daily occurring all over the country, and as the country was thinly settled in comparison with the present state of things, they could not meet so often then as now. We were repeatedly annoyed by a company of this kind, and as the summer was passing and fall approaching, they became bolder, and did not hesitate to tell anecdotes in the hearing of the audience. Preaching was then done in a little log-house. My mind was finally made up, at all hazards, to bring the disturbers in or drive them away. Rising on one occasion, announcing my text as usual, and leading off in my discourse, a clear voice fell upon our ears, relating an anecdote without. Suddenly pausing, I called their attention and respectfully invited them in, with no success. A second attempt was made, and a second failure. Finally, I told them that as telling anecdotes was fashionable and more interesting than preaching, if they would come to the door and give me a chance, and I did not beat the crowd telling an anecdote, that I would take down my sign and listen to them. The whole crowd came promptly to the door, and, as soon as they were quiet, their mouthpiece addressed me: "Parson, proceed." My proposition accepted, I was either compelled to relate an anecdote or succumb. The following occurred to me as suited to the occasion, and I related it:—

"Doubtless there are some before me who were with General Sam. Houston, eight years ago, at the battle of San Jacinto. You have doubtless heard of Tory Hill, on the opposite side of the river and in sight of the battle-ground, right near that beautiful spot occupied by the residence of President Burnett. About one hundred and fifty tories banded themselves together and sent their names to General Santa Anna with the promise that when he had whipped General

Houston and put down the rebellion, they would continue to be his loyal subjects. They assembled on Tory Hill, as the time drew near, and awaited the result of the struggle that was to decide the fate of the nation. Trembling with anxiety at every breath, those miserable, cowardly tories stood, — refugees from justice, murderers, thieves, perjurers, forgers, 'their consciences seared with a hot iron,' their eyelids smoked with the perfumes from the bottomless pit, — and with the stillness of death watching and waiting for the issue. The cavalry sallied out and brought on the dreadful attack; the artillery roared; the earth trembled. One of the black-hearted band, a little bolder than the rest, cried out, 'Hurrah for Santa Anna!' The cannon roared again. He leaped, cracked his heels together, slapped his hand and shouted, 'Hurrah for Santa Anna!' Presently there was a little recess; the cannon ceased to fire; the reports from muskets were few and far between. The hero of the band grew tremulous, and in a subdued tone said, 'Boys, what does that mean?' Down went his ear to the ground, as the contending armies receded, to catch the sound. The cannon and musket had ceased, but he distinctly heard the cracking of rifles, known to be in the hands of Texans. In a short time he put his ear to the ground the second time, and still he heard the rifles. Summoning his courage up, and with a smile of triumph upon his countenance, he said with animation, 'Boys, I'll tell you what it is, — Sam. Houston is giving it to Santa Anna. Hurrah, hurrah for Sam. Houston!' Now, gentlemen, I would not be at all surprised if that same fellow and a large part of his crowd were some trifling, back-slidden Baptists, and fallen from grace Methodists that went out regularly to preaching, and stood out and told anecdotes during the sermon, to the great annoyance of both preacher and congregation. Such men always want to be on the popular side, and are entirely destitute of principle. When it is popular to dance, drink whiskey and tell anecdotes, they are on that side; and when religion is in her silver slippers and very popular, then they are on that side."

The mouthpiece of the band of my disturbers, fearful lest I should carry off the palm of victory, interrupted and said, "Parson, how did you find that out?"

"Well, sir, I met a Mr. Smith, who was one of Sam. Houston's express bearers, and who was sent up the Trinity to tell the retreating families to stop at the Trinity River, as Santa Anna would very soon be routed and they could return to their homes. The express-man made his way back as far as Tory Hill, just in time to witness this remarkable demonstration."

During this time, all the congregation in the house and out of doors listened with the most earnest attention. After a short pause, during which no one spoke, I inquired, "Gentlemen, shall I proceed, or will you?" The leader promptly replied, "You have the floor, sir; proceed; we give it up."

Taking advantage of the victory gained, and by this time feeling more than ordinary solicitude for the cause of my great Master, I laid hold on my subject where I left it, and never in all my ministry did I have greater liberty. The little log-house did not furnish accommodation for all the congregation, but the most profound attention was given, from without as well as within, and before the sermon was closed some praised the Lord aloud, and tears flowed freely from many eyes. The victory was, under God, complete, and this was the last struggle I ever had to get the attention of the Huntsville congregation. My way, after this, was clearer than ever before, and in a very short time the organization of the church, bearing date already written, was consummated.

This church was organized upon the same principles as set forth in the Constitution and Articles of Faith of Union Association, with the addition of the following resolution: —

"*Resolved,* That any member of this church, becoming a member of any of the benevolent institutions of the day, and contributing to their support, or refusing to do so, shall be no bar to fellowship."

Without this resolution, an organization at the time could not have been made. The brethren composing the presbytery were censured for making this compromise; but subsequent developments revealed, beyond doubt, the propriety of the course adopted.

During the first year after the organization not more than one monthly meeting passed without the reception of members, either by baptism or by letter. Signs of rapid growth, peace and prosperity appeared on every hand for a time, to cheer the true friends of Christ about Huntsville. The satanic influences brought to bear upon the little flock, through the irregularities of the preachers and private members referred to, were in great measure counteracted, when a new case appeared, making our hearts bleed with shame and sorrow.

During a month's absence on my part, looking after organizations in other places, the distinguished Elder Stovall, whose acquaintance I formed while on a visit to Mississippi, in 1840, made his appearance, and preached, to the great satisfaction of the church and entire community. He was a man of very superior literary attainments, and great natural powers. My intention had already been communicated to the church to resign, and enter my first field in the State, near the Falls of the Brazos. We were all gratified with the prospect of securing the services of such a man, possessing such ability. But very soon reports reached us from Mississippi, through a reliable channel, that he not only indulged in wine, but was guilty of crimes of a baser character. While he did not visit the grocery, he drank privately, and degraded himself besides, by worshipping at the shrine of passion. He was promptly informed of the charges alleged against him. He positively denied the allegations, and promised to set himself right. In the mean time he took a school in the town. Although he received no further recognition at our hands as a preacher, he offered another impediment over which the wicked stumbled, to the great mortification of the little struggling band of disciples. He went to San Antonio, and afterwards to New Orleans, where he was hung for murder.

Surely no people has ever groaned under such burdens, imposed by such characters, more than we did in Texas, in our early history. I rejoice that I have lived to see the day when steamships, railways, and the electric wire furnish us with the means of speedy communication, so that men of this stamp cannot long remain in one locality without detection.

During the year 1844 we kept up a regular monthly appointment at a little school-house, with a dirt floor, four miles north of the present locality of Anderson, Grimes County, in the neighborhood of A. G. Perry. Here we gathered together a few Baptists, who petitioned for an organization, and on the eleventh day of November 1844, the present church at Anderson was organized by a presbytery, composed of Elders Thomas Horsely and Z. N. Morrell, with seven members. While the members lived near the school-house, we foresaw that, in consequence of the rich lands south, the centre of population would be at Fantharp's, — now Anderson, — and the church was constituted with the understanding that it would be moved there so soon as accommodations were secured.

This has been from its organization, twenty-seven years ago, one of our most prosperous and active churches. Four out of seven of the members who formed this church at its organization were living when last heard from. In the County of Grimes, two venerable sisters yet live, Sarah Kennard, and Elizabeth White, who were among the first members, who have ever been "the salt of the earth."

Seven miles west of the town of Montgomery, another monthly appointment was steadily filled, and on the twenty-fifth day of November, 1844, the Post Oak Grove church was organized, with six members. The presbytery consisted of Elders Wm. M. Tryon and Z. N. Morrell.

On the same day the organization was formed, sister Aaron Shannon, and my son, A. H. Morrell, related their experiences, and were baptized by brother Tryon. My son dated his convictions back to the revival held with the Plum Grove church, in 1839, and his conversion previous to his capture by the Mexicans in 1842.

During the latter part of 1844, and almost through the year 1845, God wonderfully blessed these three old churches, Huntsville, Anderson and Post Oak Grove. Scarcely a meeting passed, at either of these places, without the reception of members, both by letter and baptism. The population increased rapidly, and we enjoyed abounding peace and prosperity.

During this time I relied entirely upon the little churches, under God, for my support. My means were all exhausted, and I was once a little in debt. My pecuniary condition distressed me sorely; but, trusting in God, I determined to go forward with the work. It became necessary for me to visit Washington

County, and I confess the embarrassment tried me severely, when I remembered that I did not have money to pay my ferriage over the river. Rather than tell the brethren of my condition, and appeal to them for assistance, I determined to go on, in my penniless condition, and test the courtesy of the ferryman, where I had formerly been crossed free of charge. The ferryman did not happen to recognize me, and I received from his tongue some very bitter abuse. To appease his wrath I gave him my coat, as a pledge that the money should be paid on my return. With this he was quite satisfied and put me across. My real financial condition was no longer a secret, and on my arrival in the neighborhood of the Providence church, near the present locality of Chappel Hill, brother Hosea Garrett and others furnished the amount necessary to recross the river and redeem my coat.

I will here record that Providence church, Washington County, was organized with nine members, on the thirtieth day of May, 1842, by Elders W. M. Tryon, R. E. B. Baylor, Hosea Garrett and Elias Rogers. Brother Tryon served as pastor for a great number of years, and under his ministry, in 1846, there were about seventy additions by experience and baptism.

CHAPTER XIX.

ANNEXATION AND EDUCATION. — 1845.

GENERAL Sam. Houston's "Lone Star Republic," of which he had prophesied in Tennessee, and over which he had presided for two terms, making in all five years, the congress of which had been moving on wheels from one locality to another, in consequence of the repeated invasions, now had its seat of government at the old town of Washington, on the Brazos. In the fall of 1834, Anson Jones was elected as the successor of our long-admired and long-loved Houston, and the retiring president, on the ninth of December, 1844, delivered his valedictory in the town of Washington. A few extracts from this address are here inserted: —

"Gentlemen of the Senate and of the House of Representatives, and fellow-citizens: —

"This numerous and respectable assemblage of the free citizens of Texas and their representatives exhibits the best possible commentary upon the successful action and happy influence of the institutions of our country. We have met together for no purpose but that of adding another testimonial to the practicability of enlightened self-government, to witness a change of officers without the change of office, to obey the high behests of our written constitution, in good-will and fellowship, as members of the same great political family, sensible of our rights and fully understanding our duty.

"I am about to lay down the authority which my countrymen, three years since, so generously and so confidently invested me, and to return again to the ranks of my fellow-citizens. But in retiring from the high office which I have occupied to the walks of private life, I cannot forbear the expression of the cordial gratitude which inspires my bosom. The constant and unfailing support which I have had from the people in every vicissitude demands of me a candid and grateful acknowledgement of my enduring obligations. From them I have derived a sustaining influence, which has enabled me to meet the most tremendous shocks, and to pursue without faltering the course which I deemed proper for the advancement of the public interests and the security of the general welfare. I proudly confess that to the people I owe whatever of good I may have achieved by my official labors; for without the support which they so fully accorded me I could have acquired neither advantage for the republic nor satisfaction for myself. . . .

"In my retirement I take with me no animosities. If ever they existed, they are buried in the past; and I would hope that those with whom it has been my lot to come in conflict, in the discharge of my official functions, will exercise toward my acts and motives the same degree of candor.

"In leaving my station I leave the country tranquil at home, and, in effect, at peace with all nations. . . .

"Our foreign relations, as far as the United States, France, England, Holland, and some of principal States of Germany are concerned, are of the most agreeable character, and we have every assurance of their continuance.

"As to Mexico, she still maintains the attitude of nominal hostility. Instructed by experience, she might be expected to have become more reasonable; but the vain-glorious and pompous gasconade so characteristic of that nation would indicate that she is not quite ready to acknowledge the independence we have achieved. If, however, she attempts the infliction of injuries she has so often denounced, I am fully assured that the same spirit which animated the heroic men who won the liberty we now enjoy, will call to the field a yet mightier host, to avenge the wrongs we have endured, and establish beyond question our title to full dominion over all we claim.

"When I look around me, fellow-citizens, and see and know that the prospects of the republic are brightening, its resources developing, its commerce extending, and its moral influence in the community of nations increasing, my heart is filled with sensations of joy and pride. A poor and despised people a few years ago, borne down by depressing influences at home and abroad, we have risen in definace of all obstacles to a respectable place in the eyes of the world. One great nation is inviting us to a full participancy in all its privileges, and to a full community of laws and interests. Others desire our separate andependent national existence, and are ready to throw into our lap the richest gifts and favors.

"The attitude of Texas now, to my apprehension, is one of peculiar interest. The United States have spurned her twice already. Let her, therefore, maintain her position firmly as it is, and work out her own political salvation. Let her legislation proceed upon the supposition that we are to be and remain an independent people. If Texas goes begging again for admission into the United States, she will only degrade herself. They will spurn her again from their threshold, and other nations will look upon her with unmingled pity. Let Texas, therefore, maintain her position. If the United States shall open the door and ask her to come into her great family of States, you will then have other conductors, better than myself, to lead you into a union with the beloved land from which we have sprung, — the land of the broad stripes and bright stars. But let us be as we are until that opportunity is presented; and then let us go in, if at all, united in one phalanx, and sustained by the opinion of the world. . . .

"In the advance of the republic, from the earliest period of its history up to the present moment, we think we have demonstrated to the world our capacity for self-government. Among our people are to be found the intelligent and enterprising from almost every part of the globe. Though from different States, and of different habits, manners, sects and languages, they have acted with a degree of concord and unanimity almost miraculous. The world respects our position, and will sustain us by their good opintion; and it is to moral influence that we should look, as much as to the point of the bayonet or the power of cannon.

"My countrymen! Give to the rising generation instruction; establish schools everywhere among you. You will thus diffuse intelligence throughout the masses, — that great safeguard to our free institutions. Among us education confers rank and influence; ignorance is the parent of degradation. Intelligence elevates man to the highest destiny; but ignorance degrades him to slavery.

"In quitting my present position, and a second time retiring from the chief-magistracy of the republic, I feel the highest satisfaction is being able to leave my countrymen in the enjoyment of civil and religious freedom, and surrounded by many evidences of present and increasing prosperity. This happy condition is ascribable to that wise and benign Providence which has watched over our progress and conducted us to the attainment of blessings so invaluable. Let us, therefore, strive to deserve the favor of Heaven, that we may be established in all the privileges of free men, and achieve that destiny which is always accorded to the faithful pursuit of good and patriotic objects.

"It is unnecessary for me to detain you longer. I now, therefore, take leave of you, my countrymen, with the devout trust that the God who has inspired you with faithful and patriotic devotion will bless you with the choicest gifts. I shall bear with me, into the retirement in which I intend to pass the remainder of my life, the grateful and abiding recollection of your many favors."

The year 1844 passed out with the brightest prospects, as seen in this valedictory. The election of James K. Polk to the presidency of the United States was an evidence to the public mind that Texas would be annexed to the American Union at an early day. The party that nominated and elected Polk favored annexation, and the discussion of this question occupied much attention in the canvass. On the twenty-fifth day of February, 1845, the annexation act passed the congress of the United States, and on the first of March it passed the Senate. President Tyler had the honor of giving his official signature on the same day, — three days before passing out of office. This act was approved by Texas, on the twenty-third day of June, 1845.

Emigration poured rapidly out of the old States into the new one, and with the rapidly increasing population, the friends of education and religion in Texas felt that there was a mighty work before them.

In keeping with the spirit of the times the Baptists were rising up to measure arms with the educational interests of the country, and on the first of February,

1845, nearly six months previous to annexation, the charter for "Baylor University" was granted by the congress of the republic, and the institution located, where it now stands, at Independence, in Washington County. The names of the first Board of Trustees under the charter were as follows; — R.E.B. Baylor, Eli Mercer, Orien Drake, James L. Farquahar, Edward Taylor, James Huckins, James S. Lester, Robert Armistead, Aaron Shannon, Albert C. Horton, Nelson Kavanaugh, A. G. Haynes, J. G. Thomas, R. G. Jarman, and Wm. M. Tryon.

The names of the members that formed the "Education Society" that was organized at the second meeting of Union Association, in 1841, appear in a previous chapter. Through the influence of this society, and under the auspices of Union Association, this institution was chartered and placed under the care of Elder Henry L. Graves, its first president, early in 1846. This school kept pace with the progress of the country, and prospered greatly, from the very commencement. Both boys and girls were received, and recited in classes together.

In 1848, the Baptist State Convention was organized, which will be noticed at the proper place. In 1849, a committee was appointed to secure, by an act of Legislature, a change in the charter, allowing all vacancies occurring in the Board of Trustees to be filled by the convention.

And the third session of the Baptist State Convention, the committee reported that the change had been effected, and since 1850 Baylor University has been under the patronage of the convention.

In 1851 we find from the minutes of the Baptist State Convention that Elder R.C. Burleson was president of the institution, and in charge of the male department. Elder Horace Clark was in charge of the female department. At one time under this administration, we find, from the report of the board of trustees, that there were over four hundred students entered, during one year, in the two departments. Elder Burleson resigned his presidency of Baylor, and took his present position as president of Waco University, in 1861. The exact date when the two institutions were placed each under a different and separate board of trustees, I cannot give. Elder Horace Clark remained in charge of "Baylor Female College" until the summer of 1871, with only a short intermission. The Female College is at present under the care of Elder Henry L. Graves, the first president of Baylor University.

Elder G. W. Baines served for a short period of time previous to the election of Elder W.C. Crane, who has been in charge of Baylor University as president a number of years, and fills the position still.

The old "Education Society," organized in connection with Union Association, at its second session, in 1841, continued to hold its annual meetings and put forth its energies until the university was founded in 1845, and after the general interests of education were turned over to the board of trustees, this society continued to hold its anniversaries, and plead the cause of *ministerial education*. Young men, moved by the Spirit of God to enter the Christian ministry, were sought out from among the churches, and placed at Baylor University, the society promptly meeting their pecuniary liabilities. The society met with the association, in 1847, at Houston, with Elder H.L. Graves as president, and received at that meeting the sum of $305.50. In 1858, eleven years afterwards, the society met with Union Association, at the Mount Zion church, Washington County, when the Treasurer, J.W. Barnes, reported $691.74 on hand, for ministerial education. With the Bellville church, in Washington County, it held a meeting as late as 1860, and still reported on hand $211.74. The enthusiasm that prevailed at all its annual meetings was without parallel. Its friends rallied around it — and their names were legion — with their prayers and contributions, up to the late war, when its operations were suspended, because of the depressed financial condition of the country, and in consequence of the fact that the youth and manhood of the country were called to the army. It was not necessary to involve the society in debt. When God gave young men to be educated, the necessary amount could always be secured. All that the society required of its beneficiaries was, that they be approved by the churches and give evidence of their gifts. The appropriations were in accordance with the necessity that existed.

The entire expenses of some were paid, and others in part, just as their pecuniary condition required. While the brethren showed a mind to work in this glorious cause, the Lord raised up among us a number of gifted young men. During the existence of that time-honored and heaven-approved organization, no discordant elements ever entered its meetings, and I now believe, if such an organization were still hoisting its colors at every session of the Baptist State Convention, that the same spirit that animated the ''old guard'' would fire the hearts of hundreds of Baptists all over the land; and that the seven young men now at Baylor, and as many more, if God should raise them up, would be amply provided with all the means needed for their education.

The first beneficiary that entered Baylor University, under the auspicies of this society, was James H. Stribling. In 1843, he professed a hope in Christ, and was baptized by Elder Wm. M. Tryon into the fellowship of the Providence church, Washington County. In 1845, the church recognized his gift, and in the next year he entered the college at Independence

A few months previous, while on a visit to Providnece, brother H. Garrett requested me to fill his monthly appointment at Dove church, in Burleson County, of which he was the pastor. This church was organized at Caldwell, May 14th, 1843, by Elders R.E.B. Baylor and N. T. Byars, with six members. Brother Garrett served this church for several years, forty-five miles from his home.

James H. Stribling was selected to accompany me on this trip. For nearly nine years I had, with others, been praying the Lord of the harvest to raise up young men in Texas to preach the glorious gospel of the blessed God. Nearly all our ministers then at work had been borrowed from the old States. The few that had been ordained in Texas, so far as I then knew, had been converted elsewhere. Here was a youth converted in Texas, and impressed to enter upon the ministry, — the first case of the kind we met. At the Saturday meeting the young brother was put forward to open the services, notwithstanding all his objections and great timidity. We went to Caldwell expecting to stay two days, but continued a meeting of great interest for about six. The young brother was put forward at every service, and required to work. He would frequently falter, and sometimes stand with his finger pointing upwards for nearly a minute, without uttering a word. There was a deep earnestness in all he said, and the people all recognized the fact that he groaned beneath the load of his responsibility. In the midst of his hesitancy for appropriate words in which to clothe his thoughts, he did not cough, hawk, spit, nor grunt, to fill up the time. He was only waiting to express his own ideas in his own words; for there was no attempt on his part to imitate any one. During the meeting, old brother Pruitt, who will receive notice at the proper time, and I covenanted to pray to the Lord to loose the young brother's tongue. Just as we were about to close this meeting and return to our homes, brother Stribbling rose, in the midst of a weeping congregation, and asked permission to speak. God loosed his tongue on that spot, beyond all question, for about five minutes, and he has not been tongue-tied since.

After urgent solicitation on my part, he visited with me all my churches east of the Brazos, and gave evidence of rapid growth. The propriety with which he conducted himself in all classes of society was worthy of the imitation of every young man entering upon this sacred calling. Among men and women, old and young, he maintained uniformly the dignity of the Christian character. While at Anderson church, J.W. Barnes, then a Universalist, but my intimate friend, accosted me thus: "Brother Morrell, what are you carrying these boys round through the country for?" At this time brother Richard Ellis, whom we ordained on the Colorado, was with me. I replied to General Barnes, that they were good and true men; that the oldest — alluding to Ellis — was a man of usefulness; but that the younger — alluding to Stribling — was my *tart*, and that after we got him cooked he would be all right, and a credit to the denomination. Barnes replied, after a hearty laugh, "I admire your zeal, but deplore your judgement."

James H. Stribling entered Baylor University shortly afterwards, was the first beneficiary of the old "Education Society," appreciated the assistance given by devoting himself earnestly to the cultivation of his mind, till 1849, when he was ordianed, at a call from Independence church, to the full work of the gospel ministry by a presbytery composed of Elders R. E. B. Baylor, Hosea Garrett, Henry L. Graves, and J.W.D. Creath.

While he preached all he could in connection with his college life, he now entered actively upon the life of a consecrated preacher, and has never turned aside to engage in any secular employment, for a year or a month, up to the present time. The trust reposed in him by the denomination and the old "Education Society" has never been betrayed. We have ever been glad to see him rise, in any general meeting, on any occasion and at any point, and plead our common cause. Gonzales, Galveston, Wharton, and other points in the west, have long felt the power of his earnest ministry; and the church at Anderson, over which he now presides, and has for the last ten years, as pastor, still rejoices under the privileges of his ministry. All love him — none excel him.

D. B. Morrill, according to my recollection, was the third beneficiary that entered Baylor University under the auspices of the "Education Society." He was born in New York, May 17, 1825, was baptized in Michigan, February 19, 1843, and emigrated to Texas in 1845. He was found by brother James Huckins, the pastor at Galveston, driving a stage between Galveston and Velasco. His piety clearly exhibited itself at once. The watchful eye of Huckins observed that the young stage-driver carried his Bible in his pocket, and spent every moment he could with safety, even on his stage, in searching the Scriptures. He was licensed to preach by the church at Galveston on the fifth of February, 1848, and in July following entered Baylor University. The same earnestness that characterized his subsequent ministry was manifested in the prosecution of his studies. Morrill and Stribling were in the institution together for a while, and the neighborhoods adjacent to Independence enjoyed the privileges of their early ministry. In December 1851, he was ordained by a presbytery composed of Elders G.W. Baines, R.C. Burleson, R.E.B. Baylor, Henry L. Graves, and J.W.D. Creath, and entered at once upon an active, earnest ministry.

The old "Education Society" had another bright star added to its crown of rejoicing, and no reasons were ever given by Elder D.B. Morrill to cause its members to regret the means invested in his education. While Stribling went west, Morrill went east, and the churches at Montgomery, Crockett, Tyler, and Ladonia, and other intermediate points, were long fired by his zealous ministry. He was a man of small stature and presented strong indications of a feeble constitution, and yet few men have performed a greater amount of labor in the same period of time. He was eminently sound in doctrine, and hesitated not to avow his convictions. He did not deal in smooth and honeyed words when opposing the enemies of truth, but in the clearest, strongest language of which he was master fought error on every inch of ground. Some men's sermons are like beautiful round balls, arranged with so much symmetry, and presented in such elegant style, that is is hard to gather them up when the preacher disappears. D.B. Morrill's sermons were full of horns; you could not forget them if you tried, and you always found prongs to take hold of when you wanted to pick them up. The mission cause in Texas found in him an ardent friend and an able advocate. At one time serving under an eastern association as agent, with instructions to indoctrinate the churches, and at another time serving as general agent in Northern and Eastern Texas, he wielded an extensive influence. Secular engagements did not employ much of his precious time, and as he possessed the spirit of a martyr, he literally sacrificed his life in the earnest prosecution of his work.

He was a forcible writer, and was serving as associate editor of the "Texas Baptist Herald" at the time of his death.

His spirit was stirred within him, as death drew near, as he saw the advancing columns of Catholicism sweeping across the United States, and in his last affliction penned an article for the "Herald," which appeared after his death, sorely regretting the apathy of the friends of Jesus, and urging them to buckle on the armor of truth. In the prime of life he passed away, to hear the welcome, "Well done, good and faithful servant!"

I here give a private letter from sister Morrill to the editor of the "Texas Baptist Herald," and also the last communication ever penned by this noble man of God. The letter from the widow was never intended for publication; it bears upon its face the evidences of the deepest peity and her scriptural qualifications for the position she held as a minister's wife.

"Ladonia, Fannin County, Feb. 12, 1868.

"Dear Brother Link: — I enclose with this the last words of my dear husband for our paper. He wrote them on Thursday, the sixth of February, after which he told me 'his work was done; he had preached his last sermon, — he was going to a better country.' We did all we could, but God took him to rest on Monday, at two o'clock P.M. He gave us all the evidence we could ask, and even more than I ever had before, that his home was in heaven, and that he was anxious to go. He said he wished he had the world for a congregation and strength to talk; he would tell them more than he ever had before of the excellences of the gospel, and that if we were faithful to Christ we would not regret it when we come to die. This was the happiest day of his life. His work for twenty-two years had been to this end, and he wondered, if there were any infidels here, if they would not believe the testimony he could give them, and if not they would not believe though one arose from the dead.

"One moment in the arms of Jesus was better than a whole lifetime here. He blessed his children, and said we would not want. Though we were left without a home and among strangers, he felt that God would not let his family suffer. He had spent his property for the cause, as well as his life, and God would reward him. He said much more that was pleasant and gratifying to me; but perhaps I have said enough. It is some relief to me to tell it to one that I feel is doing all he can for the same good Master. I feel that all Texas sympathizes with me; but I must not weary you. Please publish these last words which I enclose, and tell the ministers that he wished them to be more faithful than they have ever been to preach the *whole* truth.

"Your sister in much affiction,
"L. Morrill."

He died on the tenth of February, 1868, and only four days before, he penned the following. —

"Ladonia, Fannin County, Texas, Feb. 6, 1868.

"To-day I am sick, and there have been but few days in several months past that my increasing bodily infirmities have not plead for rest and refreshment. But the evident shortness of my time, the quiescent and compromising spirit of many of the 'witnesses of Jesus,' and the vigorous onset of the foe, — many in number, but one in ultimate design, and *one* in the fatal tendency of their world-wide heresies, — forbid my loitering. 'The deadly wound of the beast;' kings and potentates, men of the world and all protestant sects, whose martyed sires battled nobly against the encroachments of the 'man of sin,' are upholding the black banner, and directing unconscious thousands in the course of that apocalyptic woman who made herself 'drunk with the blood of the saints.'

"The truth and the whole truth must be told. The faithful must put on the gospel armor and be rallied to the 'high places of the field.'

"But oh, what can I do? What are intellectual attainments? What are the principles of sound logic? What are matured arguments, when in the unequal conflicts with 'spiritual wickedness in high places?' Oh, thou Author of Truth, thou knowest how much I desire thy presence in every work. I desire nothing so much as to be made 'strong in the grace that is in Christ Jesus.' May thy servant ever remember that 'the race is not to the swift, nor the battle to the strong.' May I never forget that the chief end of all my labors is to win souls to Christ. May I never be tempted by flattery and the hope of gain, on the one hand, to suppress any part of thy truth, nor to falter in any duty, in view of poverty, shame, or suffering. Oh, give me faith to tread the roughest path, and to walk boldly on amid the darkest gloom. May I never seek the favor of men at the sacrifice of truth, nor ever cast one shade of reproach upon one that loves our Lord Jesus Christ."

It would require a volume of itself to record the benefits that have flowed from the "Education Society," — the founding and history of its institutions of learning, — the long list of its beneficiaries and their noble deeds in Texas, — with the long line of men and women educated there during the past twenty-seven years, who have gone forth to bless society. Over fifty preachers have received either the whole or part of their educatoin there.

The old institution chartered by the congress of the republic, still lives. From its beginning till now, streams of influence have gone out from its walls, telling all the time upon the destinies of our people. Minds have been trained there that have mastered the learned professions, and voices have been trained there that have been heard at the bar, in the political arena, from the pulpit, among the English-speaking population, among the Germans, in all the Texas Baptist Convocations, and in the Southern Baptist Convention.

CHAPTER XX.

PIONEER PREACHING. — 1846.

THE clouds that hung so heavily around that region of country adjacent to the Falls of the Brazos in 1837, when I cultivated my first crop in Texas under a guard of soldiers, and from which I was driven at an early day by the Indians, were now all brushed away by the success of our arms and the advance of civilization. The affections of my heart had always lingered around that spot, during all the years of my banishment, and early in 1846 I left the churches organized in 1845 to be cared for by other pastors, and, under a commission from the Domestic Mission Board of the Southern Baptist Convention, accepted, as missionary, the field north of the Old San Antonio road, between the Brazos and Trinity Rivers. The salary agreed upon was two hundred and fifty dollars, to be paid quarterly.

North of Anderson and Huntsville, in 1846, there was not a Baptist preacher to co-operate with me in the organization of churches, in all the vast territory lying between the Trinity and Brazos River, to the farthest limits of Texas, except Elder N. T. Byars, who then lived on Richland Creek, in Navarro County, about eighty or ninety miles from where I settled in Milam County. There was a small organization of anti-missionary Baptists in the vicinity of Springfield, with two brothers by the name of Dorsey preaching among them. Elder Byars and I met occasionally, and conferred together with reference to the interests of our Master's cause in the midst of this wide destitution.

My first trip across the country was made in January, 1846, preaching first at Leona, and then at Springfield, pursuing my old custom of asking at the close of the sermon if there were any Baptists present. This I did to make their acquaintance, and to prepare for organization. Both at Leona and Springfield I found a few Baptists, where I left appointments for February. On my second trip, without the aid of any minister, I constituted the church at Leona, on the first Sunday in February, with eight or nine members, and the second Sunday in the same month the church at Springfield, with six or seven members. Elder Byars lived in the heart of the territory now occupied by Richland Association, and in the territory now covered by the Trinity River Association there was no ministerial aid given me until the arrival of Elder R. E. B. Baylor.

Brother Baylor served as judge of the District Court under the Republic of Texas; lived first at Lagrange; resided later near Independence, where stands Baylor University, that bears his name, and around which his heart's warmest affections have always gathered. He was now re-elected under the State administration, with a district extending farther east, and embracing the counties of Leon, Limestone Falls, and Robertson. For about twenty years he held this public trust, and no man could be found that could compete with him on election day, when the popular vote was cast for District Judge. At all his courts he preached whenever opportunities were given.

On incident connected with the history of that Baptist Judge and his associate, Judge Hemphill, is worthy of record, exhibiting an independence and spirit of liberality that greatly endeared them to the people who shared the destinies of the republic.

Previous to annexation, the judges were elected by the congress, and not by the popular vote. Baylor and Hemphill had been elected, and were in the discharge of their duties. The salary of a judge was three thousand dollars per annum. The congress passed a law reducing the salary to fifteen hundred dollars. The law of course was not designed to affect the wages of the incumbents, but their successors in office.

This was a clear indication to the two noble judges, that, in consequence of the low state of the finances of the country, retrenchment was necessary. Both of them promptly went forward and resigned, announcing themselves at the same time candidates for the same office, at the reduced salary. They were both immediately elected and returned to their positions, receiving only fifteen hundred, while the other judges were receiving three thousand, dollars. Texas did not forget them in future years.

Brother Baylor came to my relief on the Colorado in 1838, as the reader well remembers, and now, in 1846, we entered the same territory. While in his official position he was rendering to Caesar the things that were Caesar's, I determined to so arrange my appointments as to catch him at our regular meetings; thus giving him a good opportunity to render unto God the things that were God's. He always arrived at the place to hold his courts on Saturday evening, and sometimes earlier. This gave us an opportunity to preach together on Sunday, and it was quite common for us to have preaching at night through the week. His preaching and Christian influence in social life contributed greatly to the cause of religion, all around his circuit. He wielded a fine influence on the bar, and our congregations at night were generally large. On one occasion I met with brother Baylor at the old Providence church, in Burleson County.

This church was organized by Elder Wm. M. Tryon in September, 1841, with twelve members. The day after the organization, brother Tryon baptized twenty persons, and during the meeting about forty were added to the church. Here lived the venerable Deacon James A. Pruitt, and the twenty converts added to this church, the day after the organization, were convicted and converted in a prayer meeting led by him without the aid of a preacher. His influence among this people to the day of his death, a number of years afterwards, reminded us very much of the ark of the convenant. Whenever he moved forward, the people moved also. Previous to the organization of Providence church, Deacon Pruitt held the prayer meetings in the yard of a venerable sister Smith, who lived at a central point in the community. She frequently pointed me and others to the stump around which anxious, penitent sinners kneeled with the old deacon for prayer.

Judge Baylor and I fell in together at this church, and there was quite an interest manifested. The church at that time was a member of the Union Association, and Elder Hosea Garrett was the pastor. He served this, in connection with Dove church, at Caldwell, several years. On one occasion during the meeting alluded to I preached, and, according to previous arrangement, brother Baylor followed me with an exhortation. He made one of his happiest efforts; but as the desired impression was not yet made upon the waiting audience, he called for brother Pruitt to come forward and address his neighbors.

The old brother ever walked humbly with God, and was willing on every occasion, as he now showed, to lend the influence of public admonition. Promptly he commended an earnest exhortation, pleading with his neighbors and their children to repent and turn from sin to God. They could stand the plain talk of Morrell and the eloquence of Baylor, but their heart-strings were soon touched by the appeals of Deacon Pruitt. The very springs in his head broke out through his eyes, and a still, small voice was heard from every quarter of reflection, and mighty contrasts came up in the minds of many sinners. "Possibly Morrell and Baylor may, after all, be working for fame; but our old neighbor is a walking epistle, read and known by us all. The book of his daily life is before us, and his conversation at every turn is seasoned with grace. Riches and earthly treasure now are not in all his thoughts." They could resist no longer, and a number bowed down, inquiring, "What must we do to be saved?" Oh, for more such deacons as this! An army of such Baptists would batter down the walls of prejudice and superstition wherever they went. Deacon Pruitt lived right on my road going to and coming from Baylor University, while my son was there at school, and it was a great privilege to be permitted to pass a night under his roof. He would light me to my bed, and then sit by me and talk the sleep out of my eyes, till the chickens crowed for day. All his conversation would be about the Redeemer and His kingdom. His was an enviable life, and I suppose that no man has ever appeared in the religious history of Texas who has attained to loftier heights in Christian life than Deacon Pruitt.

At this same meeting the case of J. G. Thomas pressed heavily upon our hearts. We were in great need of ministers, and brother Baylor, as well as myself, never lost an opportunity to give young men a word of encouragement, who were impressed to preach. On my former visit to Burleson County, in 1845, with James H. Stribling, we preached at Providence church, and learned that brother Thomas had been laboring under convictions, for some time, that it was his duty to preach, but in consequence of his great timidity it was with difficulty that we could even get him to pray in public. Men entering the ministry in Texas, under my observation, have been very much like horses. Occasionally I have met one that needed the curb and bit, in the midst of his feverish anxiety to rush forward, regardless of either propriety or consequences. Then there are others that must be coaxed along, and sometimes put under the spurs to drive them forward. Here was a man that not only had to be coaxed, but finally had to be put under whip and spur, before he could take his place in the team and do his part in dragging the heavy load along.

Deacon Pruitt, seeing the troubles under which brother Thomas groaned, brought his whole influence to bear to urge him into the work, and after a long mental struggle, under the dealings of Providence, he bowed his head and took the yoke.

Since he consented to put the harness on, he has never been found without it. The missionary spirit entered his soul in large measure, and although he has served a number of churches as pastor with much success, the most of his labors have been performed in the mission-field of Texas. In this field success has crowned his work wherever he has gaone.

During the spring of 1846 I encountered many difficulties in traversing the broad field assigned me. The Little Brazos and Navasota Rivers both had to be crossed on every trip, and there was no ferry on either stream. They were frequently swollen for weeks together, and many a time I was compelled not only to swim the main streams, but also in the low grounds adjacent. It was sixty miles to Leona, where I filled an appointment the first Sunday in every month, with both these rivers to cross. It was then, the way I travelled, fifty miles to Springfield, with several streams between, without bridges, where I preached on every second Sunday. It was forty miles to my next appointment for the third Sunday, in Navarro County, with Richland and Chambers Creeks on the way. It was one hundred miles on a direct route to my own neighborhood, where I preached in Milam County, on the fourth Sunday. With extra rides, in visiting among the scattered settlements, over three hundred miles were travelled monthly, during the entire year. High waters never prevented me from filling a single appointment.

At our first meeting after the organization at Leona, a very remarkable negro boy approached me on Saturday morning, and asked permission to join the church. Upon being asked if he believed that God for Christ's sake had forgiven his sins, he promptly answered in the affirmative; and after giving the clearest evidences of deep conviction and joyful deliverance, I told him to confer with his owner and present himself to the church. Jerry was prompt to the hour, and when an opportunity was given he presented himself for membership. In the hearing of all the congregation, he told his simple story, in a few but earnest words. There was no dream, no voice, and no miraculous manifestations rehearsed; but with plain and heartfelt utterances he convinced the congregation, already in tears, that he not only had a soul, but that his spirit had been moved by the power of God. I baptized him with some others, and very soon he expressed an anxiety to learn to read the Bible. In this he was encouraged, and in his aptness to learn soon acquired what he sought. He was granted permission and encouraged by the church to preach to his people. Up to the war, and during the struggle, he deported himself with Christian propriety, and although a quarter of a century has passed since I baptized him, he still lives to declare the good news of salvation to his race.

On this same trip, in the month of March, passing from Leona to Springfield, with only one house on the road, I found a creek swimming, about midway between these points. About two hours were lost in my efforts to head the swimming water. It was very cold and I dreaded it. Finally, my horse was plunged into the swollen stream. He swam with me to the opposite bank without any difficulty; but as he struggled amid obstructions on the Springfield side, I was compelled to dismount in the water and give the animal my assistance.

My boots were full of water, and all my clothing thoroughly saturated. A blue Texas norther whistled around my ears, and appeared almost to penetrate my quivering limbs, as I mounted the horse, at four o'clock in the evening, with twenty-five miles lying stretched between me and Springfield in a north-westerly direction, and not a single house on the way that I knew of. To my great surprise and gratification, after travelling about eight miles, my clothing now freezing, I came suddenly upon a camp by the roadside, made since my February trip. Here was a good fire, a little log cabin covered, no floors, cracks not lined and no chimney. The familiar voice of a brother in Christ was recognized; and brother Sanders, whom I had known in Washington County, and afterwards at Huntsville, invited me to share with him for the night the comforts of his camp. He had been there only a short time, had no corn for my horse, and his wagon, sent below for supplies, had not returned in consequence of high waters. It was eighteen miles yet to Springfield, with two and perhaps three dangerous streams to swim; and although the horse must shiver all night as he nipped the short spring grass, and although the missionary was informed that the family had neither bread nor meat, he decided to tarry for the night.

It was by this time almost sunset, and as I drew off my boots and exposed my wet and almost frozen feet to the fire, the good sister Sanders gave me a cup of coffee. The wind, 'tis true, whistled through the open cracks in the new log cabin, but this was far better than shivering all night alone on the bank of some swollen creek ahead, or perhaps, in the midst of my great suffering, exposing my life by trying to swim it after night. While drinking my coffee, I inquired if the landlord had guns and ammunition. This was answered in the affirmative. I asked if the dogs would tree turkeys. To this a like answer was returned. Still drinking my coffee, I ordered the guns put in good order, assuring the family that my Master had a storehouse down in the adjacent creek bottom, and that we would soon have plenty of meat.

I soon passed out of the cabin, with brother Sanders' little son and the dogs at my heels. The dogs, understanding what was wanted, preceded us into the creek bottom, same half a mile distant, and soon the fluttering turkeys were seeking protection among the trees. I was on the ground in double-quick, and a fine gobbler perched upon a limb, almost right over my head. Here I was much perplexed. The strong stick in my Master's meat-house supported the magnificent turkey well, as he stretched his long neck and sidewise turned his eye on me, uttering, "Put! put!" But the old rifle in my hand had a flint and steel lock, and, in holding the gun up in a perpendicular position, I feared, when the pan flew open, that the powder, instead of taking fire, would empty itself into my eyes. In that event the feathers would carry off the meat, and I would not be able to shoot another. But little time was given to hesitation, and taking good aim I shut both eyes and pulled the trigger. Fortunately, down came the turkey and no powder entered my eyes. By the time it was dark, we were back at the camp with several turkeys. One was immediately dressed and hung before the fire, in regular backwoods style. This was truly an earnest time for the preacher and the family.

The clothing I wore was getting a little more comfortable. But on opening my saddle-bags I found everything saturated with water from the creek I swam in the evening. My heart was very sad when I found my old Jerusalem blade and the old Concordance that I had carried for twenty-five years perfectly wet. Everything was spread before the fire, and the turkey and coffee were tasted with a sharp relish. Texans are famous for good strong coffee, and the flavor of that turkey was beyond description.

Old sister Sanders has gone to her reward, and the old brother, when last heard from, was full of age and infirmities. One year ago, while on a visit to Marlin, I met the little boy who was with me when I shot the turkey and spent a night in his family. Twenty-five years have made a great change in the country where this occurred, and have changed the boy into the head of a considerable family.

The night's rest was quite refreshing, and as the clear golden sunbeams of the morning appeared, we thanked God together for temporal and spiritual good. Preachers, I think ought always to try to make themselves useful. On this occasion I rejoiced in my ability to use a gun with success, as in other instances mentioned and to be mentioned. A visit was made to the creek, but it was not considered safe yet to cross it. A deer was killed, at this time very acceptable, and on our return to camp the wagon was in with provisions from below. Another day was spent waiting for the creeks to fall, and Springfield was reached in time for the appointment. The little church was already receiving strength by additions to the original number.

We now crossed Richland creek, and passing preached at Corsicana. Only a few small houses there then. Chambers' creek was crossed and an appointment filled in a settlement beyond, where the Providence church was organized in July, 1846, without any minister present but myself, with fourteen members. People from a long distance off met us there in March and afterwards, and we had a large congregation for that day, which met us regularly.

As I retraced my steps homeward, I turned up Richland creek, and visited brother N. T. Byars, preaching in his community. According to my recollection, he had some time previous to this organized the church known as Society Hill. Sometimes on my return I preached at Springfield. On my arrival at the little Brazos the weather was quite cold, the canoe on the opposite side, and nobody in sight or hearing. Here was another large swollen stream to swim. Reaching home, I filled my appointment where the Little River church was organized in 1847, with six white and one colored member, by Z. N. Morrell.

You now have the history of one month's work performed by a pioneer missionary. A change of horses was required every trip, and these trips were made in succession, monthly, for two years, making over seven thousand miles. My salary each year was two hundred and fifty dollars, and the last year I spent three hundred dollars to keep myself in horses.

In the fall of 1846 the old Union Association received into her bounds a valuable addition of ministers. No ship that ever ploughed the waves between New Orleans and Galveston, I suppose, ever brought at one time a more valuable cargo for Texas than the one that landed Elders J. W. D. Creath, P. B. Chandler, Henry L. Graves and Noah Hill. Elder Graves came, under a call from the Board of Trustees of Baylor University, to take his position as president of the school. The others came under appointment of the Board of the Southern Baptist Convention, as missionaries. Over a quarter of a century has passed, and three of the number still live.

Elder Noah Hill passed to his reward on high in the fall of 1867, having faithfully devoted twenty-one years of the prime of his life in Texas to the cause of the Master whom he loved much. I formed his acquaintance first at the session of Union Association, held with the church at Houston in 1847, and met him frequently afterwards. Many a time have the fires of my soul burned under the strains of his pious eloquence.

He was born in Rutherford County, North Carolina, in 1811, and was baptized in 1837. He was married in early life, and with a wife and three children he moved to Mercer University, and there acquired an education. Largely imbued with the mission spirit, he accepted the appointment alluded to, and coming to Texas, settled first at Matagorda, preaching much in the adjacent country. In the midst of the storm that swept over that beautiful city on our coast, he suffered much loss and moved into the interior. Washington and Fayette Counties long felt the power of this earnest preacher, while he served as pastor both at Lagrange and Brenham.

As a preacher, his zeal and earnestness, upheld by a large benevolence, were felt by all men and congregations with whom he came in contact. His head was full of waters, and his eyes a fountain of tears, that wept freely over lost sinners. Those who heard him in the midst of his masterly exhortations can never forget him. He was sound in doctrine and pressed his conclusions with power. Large numbers repented under his appeals. He preached to the heart.

Modest and unassuming in his manners, he never sought position in any of our general meetings. Seldom could he be induced to make speeches, even on questions of great interest. In the pulpit he was always at home, and the people heard him gladly. The effect produced on my mind under two of his efforts will never be forgotten. One was made before the Little River Association, at Lexington, Burleson County, and the other before the Baptist State Convention at Anderson. On the former occasion he preached on "faith, hope and charity," and as he showed the successive steps of the believer, from the moment trust in Jesus possessed his soul, until faith was lost in sight, hope in glad fruition, and he entered the boundless sea of everlasting love in heaven, our souls leaped for joy that was "inexpressible and full of glory." On the latter occasion he pleaded the mission cause on Sunday, at eleven o'clock, before the Baptist State Convention, in which he showed himself master of the glorious theme. Towards the close of his life he suffered under great bodily afflictions. A sermon in his memory was preached by Elder R. C. Burleson to the congregation at Brenham, in 1868, where he held his last pastorate. The text suggests the power of his influence: "He being dead yet speaketh."

CHAPTER XXI.

THE RACE TRACK AT SPRINGFIELD. — 1846.

WHEN I indicated to the brethren and friends in my old field, in Union Association, my intention to occupy the field now embraced in Trinity River and Richland Associations, they protested against it; some of them declaring that life would be hazarded among cut-throats and desperadoes. But I never did have any fears in going into any territory occupied by people speaking my mother-tongue. There is something in the name of Jesus that will overcome prejudice and produce respect for the minister who, with discretion, declares it faithfully; and with kindness, moderation, and firmness, a victory may be gained over our worst enemies.

Springfield, in 1846, was the head-centre for a large number of gamblers and professional racers. Some of these were desperate men, and with the blood of their fellows upon their hands. Some fifty families lived in and around the little village. There stood the grog-shop, — the enemy to peace and good society, to say nothing of Christianity. I rarely ever got far enough on the frontier to avoid these sinks of sin. God have mercy upon the man that measures liquid fire to his neighbor by the quart! He may be recognized in decent society here, but the terrors of the second death hang upon his future pathway.

Near the town was a race-track on which much labor had been expanded, and from many miles, in almost every direction, at certain seasons of the year, sportsmen brought trained horses to this track, either to win or lose large sums of money. The characters that hung round the old town of Washington in 1837, of whom I have written, and the disturbers that tried me sorely at Huntsville in 1845, had taught me some important lessons, while I taught them and others the way of life. If you want to get a victory over animal or man, the best policy is not to let him get the start of you. This lesson, I take it for granted, was taught by the Saviour, when he said, "Be wise as serpents." These characters I watched with a jealous eye, on my new field of labor, intending to do them all the good I could; but at the same time determined, in answer to prayer, that my Master should get the victory. My former conflicts caused me to whet my ingenuity to keep a keen edge, in the event of an issue.

As the spring advanced and summer drew nigh, the ordinance of baptism was demanded for the first time at Springfield, and administered. The people "from all the regions round about" came to witness it. The races were on hand for the following week, and gamblers, from near and from far, were on the ground.

The baptism was over, and the time drawing near for preaching. The gamblers, about fifty in number, retired to the grocery to get a drink. I was having more trouble than usual in getting my mind centred upon a text and a subject. A little season was given to prayer, under the hill hard by; but as I walked towards the court-house where we preached, no light appeared. Preachers who read this will know how I felt. As I passed close by the grocery, on my way, I observed that the door was shut, and the fifty sportsmen followed after me. This was more than I expected, but it afforded much gratification. The reflection passed through my mind that these men would not pray for me, and my spirit was somewhat cast down. An expression from Paul now fastened itself upon my mind, and a train of thought, formerly digested, came to my relief. Soon all were quietly seated, the sportsmen by themselves, and, after the ususal preliminaries, the text was announced: —

"Wherefore, seeing we also are compassed about with so great a cloud of witnesses, let us lay aside every weight and the sin which doth so easily beset us, and let us run with patience the race that is set before us, looking unto Jesus, the author and finisher of our faith." — Hebrews xii. 1, 2.

"Religion is true. A cloud, a great cloud of witnesses hovers about us, bearing testimony to its divine authenticity. Man's own conscience, in the light of revealed truth, bears witness to the fact that he is a sinner, estranged from God, and deserving punishment. The claim of religion are pressed upon him, in this helpless state, designed and able to bring him back to God. By nature he is rebellious, and, according to Scripture, an unbeliever.

"As the jury, a short time past, was impannelled in that box (pointing to my right), and sworn to decide, according to law and testimony, the guilt or innocence of the man who was tried for his life, so, at this hour, you may consider yourselves sworn in before God, the mighty Judge, to decide your guilt or innocence, and, at the same time, which *race* you will run to-morrow.

"First: when the eye of man, in ancient days, beheld the heavens spread out like a curtain, and dotted with millions of blazing orbs of light, God declared, by the mouth of his servant David, in the nineteenth Psalm, that 'The heavens declare the glory of God, and the firmament showeth his handywork.' He further declared by the mouth of the great apostle, who was caught up to the third heaven, that 'The invisible things of him from the creation of the world are clearly seen, being understood by the things that are made, even his eternal power and Godhead; so that they are without excuse.' — Romans i. 20.

"The testimony given by the creation and these glorious manifestations certainly establishes the fact in the mind of every juror, that there is a God, a mighty God, the father of us all. None but a fool will return the verdict, 'that there is no God.'

"The second point taken in this case is that part of the testimony rendered by the Word of God. The witness is full of age, and clearly testifies what he saw and heard. Long before the morning stars appeared, or even the sons of God shouted for joy, this 'Word was with God, and the Word was God.' His testimony was never doubted in heaven, among the angelic throng who ever sing his praise. Early in the history of man, he made known the promise that the seed of the woman should bruise the head of the serpent. More than seven hundred years before the shouting angel host declared the presence of the new-born King, Isiah saw, with prophetic eye, the scenes transpiring in Bethlehem, and wrote, 'Unto us a child is born, unto us a son is given; and the government shall be upon his shoulder; and his name shall be called Wonderful, Counselor, The Mighty God, The Everlasting Father, The Prince of Peace.' The prophet Micah, hundreds of years before, was told the place of his birth and the character of the personage. 'But thou, Bethlehem Ephratah, though thou be little among the thousands of Judah, yet out of thee shall he come forth unto me that is to be ruler in Israel; whose goings forth have been from of old from everlasting.' The birth, the life, the cruel death, and triumphant resurrection of the Son of God, are facts clearly set forth under the testimony of our second witness, and he who doubts must be damned, 'because he hath not believed in the name of the only-begotten Son of God.'

"The third witness on the stand is the *Spirit* of God. The prophets, long before, foretold his coming. Jesus while on earth made promise, that when he left the Father would send 'the Spirit of truth.' On the day of Pentecost he came with 'a sound from heaven, as of a rushing mighty wind, and filled all the house where they (the disciples) were sitting.' His testimony quickens, enlightens, sanctifies, comforts and saves. He convicts of sin, and faithfully points to the remedy, Jesus, who is both the priest and sacrifice.

"In the fourth place, I bid you pause, and consider well the testimony given by the long line of witnesses, — from Abel, who 'offered unto God a more excellent sacrifice than Cain, by which he obtained witness that he was righteous, God testifying of his gifts;' to the patriarch Abraham, who 'believed God, and it was imputed unto him for righteousness; —' and to the list of saints, under both dispensations, who have testified that God has power on earth to forgive sins.

"Men were trianed in ancient times, in the East, both to fight and run. Before entering upon the track to run for the prize, they were required to spend months under the most rigid rules of diet and exercise, and then in the race every garment that would offer the least obstruction was laid aside.

"In connection with oriental customs, let us consider, on this occasion, some of the usages of modern racing, that may serve us in illustrating the Christian's flight along the race-track of life, for the glorious prize and crown of rejoicing that lie in full view of the eye of faith, at the end of his race. Before me I see quite a number of gentlemen, of all ages, from the youth in his teens to the man past the meridian of life, who expect to enter their horses upon that beautiful track on the prairie, about half a mile from where I stand, towards the rising sun, to-morrow morning at ten o'clock. If I make mistakes in stating the preparations required and the laws that govern the track, there are gentlemen here who can and who have permission to correct me.

"Men, in ancient times, were frequently trained, like your horses now, to run with weights on the feet, and sometimes heavy burdens on the back. Relieved of these weights, the animal feels greatly encouraged and bounds eagerly away. God sometimes allows grievous burdens and temptations to be placed upon his people, who run for the prize, and by suddenly removing these allows them rapidly to ascend the shining way. Men were not permitted, among the ancients, the use of wine or any strong drink, during the months of training. It enfeebled the muscles, and if indulged in on the day of the race made the man reel upon the track, and perhaps brought him in collision with his opponent. Hence the use of wine and strong drink was strictly forbidden. Large quantities of food, highly-seasoned, have been denied your horses now for several days, as every experienced racer among you knows that full-fed horses are both sluggish and clumsy. Temperance and moderation are lessons everywhere inculated for the government of those who would successfully run the Christian race. The indulgence of every wicked passion stupefies and retards the man of God in the pursuit of spiritual ends.

"Among civilized nations there are rules and customs in war that the honorable never violate. So among sportsmen there are stipulations and well-known regulations, to which the honorable among your number rigidly adhere. He who would cheat and defraud hesitates not to tax the ingenuity of his deceit behind the curtain, and, with a lie in his heart, sign the rules governing an honorable race. Honest men would spurn the suggestion of 'cross-balancing' a competor's horse.

"Lest some of the audience may not understand the phase, I will suggest that the smith employed to shoe the horse is sometines bribed. The shoes are made of unequal weight by inserting in one a plate of lead. This is done with so much ingenuity that the owner is not at all likely to discover it. The horse in this condition is greatly hindered.

"The honest racer would on no account allow the horse that was relied on to win his thousands to be beaten by moonshine, in the presence of a few witnesses, that upon this information others might be led into a snare and lose their money. (Here there was some uneasiness and elbowing. In a low tone, one said, 'He has been there and knows all about it.') In my boyhood days, in Tennessee, on several occasions, I rode upon the track and got an insight into some of the tricks of the swindlers. (A voice, 'I told you so.') Miserable fellows they are. Among the ancient, honorable racer, such characters, when detected, were ruled out as perjured villans; and when men are guilty of these things now they ought to be ruled out and never allowed to enter again. 'Provide things honest in the light of men' is a scriptural injunction; and the man, professing Christianity, who gets position or riches at the expense of his neighbor's welfare, deserves to be ruled out of the Christian race.

"The text describes a spiritual race, given the qualifications for it, and the weight that must be thrown aside. We all, as Christians, have vulnerable parts. These weaker defenses in human character Satan well knows, and against these he makes his strong assualts. These attacks lead us into the besetting sin. 'The sin that doth so easily beset us' must be ferreted out and studiously held in check. Thus we lay aside the weight.

"We inquire first for the preacher's weight. It may be his want of faith in God. Difficulties, like dark clouds, hang upon his pathway, driving away, to all human appearances, prospects of success. Enemies to religion, deceivers, and those who corrupt the youth of the country, have influences where and when he seems to have none. Sacrifices and dangers hang upon his pathway. In the midst of his trouble, and in the absence of that strong faith in God that rises above the clouds, Satan whispers, 'All the kingdoms of the world and the glory of them are mine.' 'All these things will I give thee, if thou wilt fall down and worship me.'

"Pride sometimes enters the preacher's heart, and the devil sits by him in the pulpit, tells him of the great number of souls he has been the honored instrument in converting, tells him of the powerful effects of his sermons, and whispers flattery and lies in his 'itching ears' until he is almost beside himself with vanity. The devil knows that God despises vanity, and hence he is always glad of a chance to stuff a preacher full of it. Full of this, he labors to please men; and Paul declared, 'If I yet please men, then would I not be the servant of Christ.'

'The preacher must have 'his loins girt about with truth.' When Satan can get a man into the pulpit preaching traditions, 'teaching for doctrines the commandments of men,' he regards him and those who follow him an easy prey. Without the girdle of truth, which is the word of God, the preacher is weak in the back, and certainly unfit for a race. The truth is powerful. An old soldier said, 'The truth will stand alone as soon as it is born, but falsehood will fall with all the props that men can place around it.' A minister should never take the truth to knock error down. Truth should simply be placed by the side of error. As light appears, darkness hides itself. As Dagon fell in the presence of the Ark, so error melts away in the presence of truth.

"What is true of the minister, as already stated, applies with force to deacons, appointed as officers in the church of God. If in the exercise of their office faithfulness is wanting at any point, and they do not rule well over the department under their special care, the devil wields, through their neglect, a mighty influence. The support of the pastor, expense of keeping the house of God in good order, and the care of the poor of the flock, are all placed on them. They are only required to pay an individual part; but they are full of weight and disqualified for the race as long as they have any disposition to surrender, short of the accomplishments of their work.

"We now pass to the multitude asking permission to enter this race. This class embraces the king on his throne, and the peasant with his hoe. While all are included, the same qualifications are required of all.

"First, as these are all sinners, 'dead in trespasses and sins,' blind and under the power of darkness, they must be made alive, and sight given instead of blindness. Dead men cannot walk, and blind men cannot see to run for a prize. Quickened by the power of the Spirit of God, the hand of faith takes hold on truth, and the eye of faith sees 'Jesus, who is the author and finisher of our faith.'

"The second step to take is the oath of allegiance to him who instituted the Christian race. Jesus is the author and finisher. As he was buried and rose agian, he requires that his disciples shall testify, by their burial in baptism, and their resurrection from the watery grave, their belief in him. Determined to 'walk in newness of life,' their sins must be laid aside, and the love of them; and 'forgetting the things that are behind,' they are fully prepared to enter the race, with immortality and eternal life as the prize offered by Christ at the end. If you are not prepared to run this race, clear the track, and dare not throw an obstacle in our way.

"In the midst of the confusion and excitement in arranging for the race, men sometimes fail to be ready at the set time. Sometimes the stakes agreed upon are not there, and again the horse and his rider won't be controlled and submit to the rules governing the race. In either case your judges rule them out, and demand a forfeit for the failure. The Judge who decides the qualifications for the spiritual race cannot be deceived. Those demanding entrance must possess the legal qualifications, or be driven away with the payment of a fearful forfeit. To-morrow morning, ten o'clock, will find some of you unpreprared for the race on the Springfield track, and in the midst of your high anticipations, indulged in for months, the disappointment and mortification will be very great, when, under the regulations, the forfeit of a hundred or a thousand dollars is demanded. The forfeit that is demanded of all who fail to make due and timely preparation for this spiritual race is not so easily paid. In the payment of this forfeit, the deathless spirit must take hold of terrors that hang around the second death. A failure to enter is the forfeit of soul, body, peace and happiness forever.

"The 'Prince of the power of the air' furnishes a number of horses for the race-track of life, and promises the rider that he shall win happiness as the prize.

Riches, honor, pleasure and passion are silver-mounted steeds, and Satan asserts that they are swift as the wind. Although they have failed, in a thousand instances, to win the price that was promised, allured by the beauty of the horse, and deceived by the promises of his Satanic majesty, men continue, in large numbers, to hazard everything upon their speed.

"By the side of these trained steeds, I'll enter, to-day, the *white horse* of the *gospel*, and *stake my all, body, soul, and spirit,* — my interests for time and eternity, — upon the race, with happiness, immortality and eternal life, to be reached at the end. For over five thousand years men have been risking their all upon this horse, and as yet he never has lost a race. Wicked men, it is true, once crucified and slew him, and rejoiced for a season over his downfall, as he slept in the grave; but he had power to lay down his life, and power to take it up again. On the third bright morn, mocking the pretensions of the devil and his minions, he chained death to his chariot wheel, and swifter than the wind, when forty days had passed, he sped his way to the heavenly court, and grasped the prize, bearing with him the first fruits of the resurrection. As he entered the avenue that revealed his approach to the gates of the eternal city, the shout of the angel hosts reverberated along the vaults of heaven, 'Lift up your heads, O ye gates; and be ye lifted up, ye everlasting doors; and the King of glory shall come in.' He has 'led captivity captive' and 'received gifts for men.'

"Around me to-day are a few who have counted up the cost, and equipped, as the laws governing this spiritual race direct, are running, with the flight of time, 'the race that is set before us, looking unto Jesus, the author and finisher of our faith; who, for the joy that was set before him, endured the cross, despising the shame, and is set down at the right hand of the throne of God.' In this wilderness of Texas we are determined, under God, to do all that we can to lay broad and deep the foundations of society upon gospel principles, and hope, through grace, at last to win the prize. And now, I warn all, that he who runs on any other track than that which is stained with Jesus' blood loses everything and gains nothing."

. . . .

I will not affirm that this was the sermon, verbatim, preached at Springfield. No notes were ever made of it, except in my memory; but impressions were made upon my mind that are vivid still, and I know that this is the substance.

I left the place greatly mortified, and cast down in my feelings, believing that I had lowered the dignity of the pulpit; and that instead of magnifying my office and inspiring respect for the ministers of Jesus, I had furnished grounds for jesting and merriment on the race-ground the following week. I tried to forget it, and erase the sermon and the occasion that called it forth from the book of my recollection. The greater the effort, the more signal the failure. My troubles and misgivings relative to it but served to burn it the more deeply upon my mind.

One thing I observed: at my subsequent appointments at the place, marked respect and attention were given from the local members of this club; but this manifestation never justified me, in my own estimation, for using in the pulpit the phrases of racers and gamblers, and thereby revealing to the people at large my familiarity with the conversation and customs of the profession.

Fourteen years afterwards I attended a session of the Trinity River Association, and there was accosted, by a warm-hearted, intelligent gentleman, with the appellation of *brother.* He shook my hand coridally, and, seeing I did not recognize him, reminded me of the sermon to the racers, in 1846, when, as he said, I bet on the *white horse* of the gospel. He repeated the text, and, after stating some of the more prominent positions taken in the sermon, said, "I was one of that crowd of racers. I had a white horse on the track the following week. The sermon was common talk during the races, and every time I looked at my white horse I was forcibly reminded of the white horse of the gospel. After I left that place God gave me no rest till peace was given by faith in Christ." He had been a member of Trinity River Association. What was true in his case, he said was true in the case of a number of others of the same crowd. This was the first comfort since the sermon was delivered.

It is fifty years since I entered the ministry, and, after mature deliberation, I here record my conviction that there are too many strait-jacket, tight-laced, cut-and-dried sermons delivered. I mean by this sermons are prepared for certain hours, and delivered irrespective of the surroundings at the time. A sermon may be good on one occasion that is entirely inappropriate for another. Subjects and sermons ought to be studied, and studied profoundly; but if the preacher finds that the congregation is composed of sinners, when he was expecting saints, then he had better lay aside his sermon on election and preach on repentance, even at the risk of criticism. Subjects well chosen make half the battle, and Satan trembles when the preacher takes good aim.

S. A. Hayden, D.D., L.L.D.

Rev. Z. C. Taylor

Rev. W. R. Maxwell

O. C. Pope, D.D.

Rev. S. L. Morris

R. E. B. Baylor

Rev. W. B. Bagby and Wife

Rev. James Huckins

Mrs. Myra E. Graves

Rev. N. T. Byars

Rev. A. T. Hawthorne

J. H. Luther, D.D.

Rev. A. J. Holt

J. H. Stribling, D.D.

W. A. Dunn

R. C. Buckner, D.D.

C. C. Chaplin, D.D.

R. B. Burleson, D.D., L.L.D.

H. L. Graves, D.D.

R. C. Burleson, D.D.

Rev. J. W. D. Creath

Rev. G. W. Baines

Jonas Johnston, D.D.

Rev. M. V. Smith

S. J. Anderson, D.D.

F. M. Law, D.D.

B. H. Carroll, D.D.

W. C. Crane, D.D., L.L.D.

J. B. Link, D.D., L.L.D.

R. J. Sledge

Rev. W. D. Powell

CHAPTER XXII.

TWO ASSOCIATIONS. — 1847

NO year, in the early history of Texas Baptists, dawned upon us with brighter prospects than 1847. From the south came news of the most cheering character. Huckins at Galveston, Tryon at Houston, and Hill at Matagorda, were receiving and reporting evidences of divine favor. Chandler was at Lagrange, and Creath at Huntsville, realizing those blessings that attended the missionaries of the Marion Board so long in Texas. Henry L. Graves, the first president of Baylor University, already was enjoying an earnest of that success which rested upon the beloved institution during the years that followed. Hosea Garrett, in Washington County, and R. E. B. Baylor, on his district as judge, preaching wherever he went, both rejoiced on their respective fields; while N. T. Byars and Z. N. Morrell were, under God, laying the foundations of the Trinity River Association.

East of the Trinity River, churches were springing up in different sections that were soon to be banded together in associations; and from every quarter of the State, wherever the gospel was preached, God poured out his Spirit and revived his work. Previously the strength that gathered around the infant churches was principally from immigration; but now large accessions were being received by baptism, and anticipations formerly indulged were being realized. 1846 laid the foundation for blessings in large measure realized during the years that followed. Ascending the hill to that beautiful spot occupied by Baylor University, which presented then and does still a scenery in every direction of surpassing grandeur, and looking across the prairies stretching every way, north, east, south, and west, I frequently, in 1847, held up my head and rejoiced that the wilderness was blooming as the rose.

At the eighth annual session of the Union Association, in 1847, the churches at Lagrange, Macedonia, Plum Grove, Bethany, and Rocky Creek, petitioned for letters of dismission, having in view the organization of a new association. The letters were granted, and a committee, consisting of Elders Garrett,

Huckins, Tryon, Morrell, and brother J. M. Hill, was appointed to aid the churches in forming the new organization.

Delegates from nine churches, representing one hundred and nineteen Baptists, met with the Rocky Creek church, in Lavaca County, on Thursday before the third Sunday in November, 1847, and, after a sermon from Elder Z.N. Morrell, called Elder Hosea Garrett to the chair. So soon as the letters were read the convention was organized by electing Elder Noah Hill, president, and T. J. Pilgrim, Recording Secretary. Articles of faith were read and adopted, and the first article of their Constitution declared that "This association shall be called the *Colorado Baptist Association.*"

The association was then organized, by electing Elder Richard Ellis, moderator, and Thomas J. Pilgrim, clerk. The churches composing the body were among the first organized in the state. Six counties were represented by the nine churches, in the territory reached from the city of Austin to the coast, and as far west as the Guadalupe River.

The first moderator of the Colorado Association was Richard Ellis, a Virginian by birth. He came to Texas in 1837, at the age of twenty-four, and solicited employment from me, in the old town of Washington, on his arrival in the State. He was a young man of enterprise and great industry, giving evidence of that perseverance that marked the man in his ministerial work. We have sometimes been led to doubt the conversion of men, who claimed to be Christians, in the midst of laziness. A man that can't be cured of this disease had better follow something else than preaching. We had no saw-mills in Texas then, and being in need of lumber, brother Ellis seized the whip-saw and rendered me some valuable assistance. He was a young man of great piety, and at the time he was with me, in 1837, was under impressions to preach. The most, however, that we could induce him to do, was to pray in public, which he did in the midst of the scoffers at Washington, described in a previous chapter. He moved to the Colorado, and became a member of the Macedonia church, some eight or ten miles below Austin. This church was organized in May, 1841, by Elders R. E. B. Baylor and J. Woodruff.

By this church he was licensed, and as there were no ministers at hand to ordain him, and as none could visit them at the time, he was sent down the river to Plum Grove, and set apart to his work, in the presence of that church, by Elders R. E. B. Baylor and Z. N. Morrell, in 1842.

Brother Ellis was a man of strong mind, with only a medium English education; yet his words were always acceptable, and his speech of great power. Zealous in the great cause he had espoused, he devoted to it all the energies of his manhood; and in the pulpit, every movement of his body, flash of his eye, and utterance of his tongue, revealed the soul of earnestness. A number of communities, in the bounds of the Colorado Association, felt the power of his ministry, and as the rising man among the churches, he was placed in the moderator's chair.

He was a man of great usefulness; but his ministry was of short duration. He exerted himself powerfully in his sermons, and many of us suppose that he passed to the grave prematurely in consequence of it. The date and circumstances of his death are not before me.

Thomas J. Pilgrim, the modest, unassuming, intelligent, pious, model Christian gentleman, was the first clerk of the Colorado Association. He was in Texas in 1829, and the superintendent of a Sunday school at San Filipe, of thirty-two scholars, the first organized in Texas. He has been appropriately styled the Sunday-school man of Texas, not only because he founded the first, but because he has contributed all in his power to promote the Sunday-school cause, during all these forty-three years past. The church at Gonzales has long enjoyed the counsels and personal influence of this faithful servant, who was still living when last heard from.

The second session of the Colorado Association was held in the city of Austin in 1848. The church at Austin was organized by Elder R. H. Taliaferro, in July, 1847, with eight members. He was their pastor at the time the association convened with the church, and with only short intermissions has filled the position up to the present time.

Brother Taliaferro came to Texas, as a missionary, early in 1847, and located at Austin in February, previous to the organization of the church in July. The Colorado valley has long been blest with the ministry of this eloquent preacher. Simple in his manners and conversation, making the little child rejoice in his presence, he may justly be styled the Apollos of the Texas churches. At all our general convocations, which he attended regularly, previous to the war, none who heard him forgot the power of his imagination and the passion of his eloquence.

He has always been a close student, and as a writer has few superiors. He filled for some time the office of associate editor of the Texas "Baptist Herald," and hundreds of Baptists eagerly read every article that appeared over his signature.

The churches composing this body were greatly blessed. They had an efficient ministry, and, as the resources of Western Texas developed, churches sprang into existence. Accessions of new churches appear in the minutes of every meeting, and from nine churches, in 1847, with one hundred and nineteen members, the number continued to increase until 1858, when the minutes show forty-six churches, with a membership of nineteen hundred and seventy-two.

Elder P. B. Chandler was chosen moderator, in 1849, at its third session, and was honored with this position frequently afterwards. Brother Chandler came to Texas, as a missionary appointed by the Domestic Mission Board of the Southern Baptist Convention, in the fall of 1846, and located at Lagrange, Fayette County.

He has ever been consistent both in theory and practice. Sound in the faith at all times, he has always allowed his moderation to be known unto all men. He has been very useful among his churches and ever enjoyed their confidence. He has never appeared as a shooting, blazing comet, now here and then yonder, sometimes flashing and again invisible, but has steadily maintained his hold on the affections of his people and quietly toiled on among them in word and doctrine.

The leading spirits in this body have always been mission men. No disorganizing, anti-missionary element ever embarrassed the aggressive policy of the Colorado Association. Those who moulded public sentiment in this body from the beginning were themselves missionaries, and in active sympathy with the Boards in the east. With such men as Hill, Taliaferro, Chandler and Ellis, active and yet acting in perfect harmony, nothing but success might have been expected. In 1850, after receiving some aid from the Baptist State Convention, the association was reaping the benefits of one half the time of Elder J. H. Stribling, engaged as missionary. The following extract from his report at the fourth session shows something of his work and the vast field to be occupied: —

"Since my appointment until this date I have travelled one thousand four hundred and sixty miles; preached fifty-five sermons, and delivered twenty-three lectures and exhortations. I have visited and preached at the towns of Wharton, Matagorda, Texana, Port Lavaca, Indianola, Petersburg, Halletsville, Seguin, Gonzales, and other destitute portions of the country.

"Multiplied scores around us are calling for some one to break to them the bread of life. Our contiguity to Mexico, with its numerous multitudes oppressed by superstition and moral degradation, gives us facilities to improve their civil and religious condition not formerly enjoyed. Our German population is in a condition scarcely less deplorable. Nor should we neglect the important duty of watering where we have already planted in the Lord's vineyard.

"Our encouragements for increased energy and devotion are great. Weak and inadequate as our efforts have been, we have much cause for gratitude to God. Hitherto the Lord has blessed us. Already has the desert waste begun to blossom as the rose, some fruit has been manifest, precious souls have been redeemed, light is spreading, darkness is receding, and the leaven of truth is spreading through the minds of our population."

At the fifth session, held with the Macedonia chhurch in 1851, we find a report from the Executive Board, to the effect that Elder J. A. Kimball was laboring as an evangelist west of the Guadalupe River.

At what time brother Kimball came to Texas I am not able to state definitely, but my impression is that he had not been in the State long previous to 1851. He did the work of an evangelist faithfully, and served as pastor at Seguin, Wharton, Bastrop, and other points west, in the early part of his Texas ministry. He has done faithful service among the churches in Union Association. Brother Kimball has done much to indoctrinate our churches and stimulate our people to search for knowledge.

He is not what the world at large would call an orator, but his large fund of information on all subjects with which he comes in contact commands the attention and respect of his audience. His great excellence is in his pen. This he wields with a master-hand, and has, during the years that are passed, added much interest both to the old Texas Baptist, and since the war to the Texas "Baptist Herald." Some may say of him, as they said of Paul, that "his bodily presence is weak;" others may say that "his speech is contemptible;" but all who read after him must say that "his letters are weighty and powerful."

Throughout the vast territory over which the Colorado Association sent her missionaries, churches have continued to spring up in large numbers, as will be seen in giving notice of new associations organized in the west in later years.

While the Baptists in the west, in 1847, were a unit on doctrine, and acting in concert under the articles of faith published in the "Encyclopaedia of Religious Knowledge;" while the necessity that gave rise to the organization of the Colorado Association was that the territory was too large for one body successfully to prosecute its work; and while peace and prosperity abounded among the western churches, the brethren in the east were in trouble. The missionaries and anti-missionaries, instead of harmonizing, had been gradually defining the lines of disagreement since the organization of the Sabine Association in 1843. Elders Isaac Reed and Lemuel Herrin, who co-operated in the formation of this body, could no longer consistently act in concert. Reed, and those who acted with him, violently opposed all mission organizations. Herrin, and those acting with him, although in the minority, could no longer consent to remain in a body that opposed mission enterprises, Boards, Bible Societies, and, in a word, all benevolent organizations outside of the church.

The minutes of the Eastern Missionary Baptist Association, now before me, have the following, as the first item: "By agreement, the representatives from Macedonia, Henderson, Eight-mile and Border churches, met with the Border church, on Friday before the first Lord's day in December, for the purpose of forming an association."

The convention thus assembled in December, 1847, adopted the Articles of Faith contained in the "Encyclopedia of Religious Knowledge," and adopted the following resolution: —

"*Resolved,* That it is due to our brethren of the Sabine Baptist Association, to the community at large, and to ourselves, that this association should state frankly to the world the reasons which induced us to separate from the Sabine Baptist Association, and to organize the Eastern Missionary Baptist Association, viz.: That the Sabine Baptist Association, at its last meeting, refused to sanction the doctrines of the annexed circular, and declared a non-concurrence with its principles." The theme of the circular letter referred to was, "The Strength ot Christian Charity," and the essay is in spirit and letter a missionary document.

The old Border church, with which the first meeting was held, was in the organization of the Sabine Association, and, as we have before recorded, was organized, by Elders Reed and Herrin, in 1843, with eight members. Its location was in Harrison County.

The Eight-mile church was organized in 1845, in Harrison County, by Elder Lemuel Herrin, with five members.

The Macedonia church, Panola County, was organized with thirteen members, by Elders Reed and Herrin, in 1844.

The date of the organization of the Henderson church, Rusk County, and the presbytery by whom it was constituted, are not in my possession. The minutes of 1848 show that its number was seventeen in 1847.

Elder Herrin was the first moderator, and J. B. Webster clerk. The second session was appointed for September, 1848, to be held with the Macedonia church, and Elder D. Lewis was requested to preach the next introductory sermon. A short time after his brother Lewis paid us a visit, and while in Leon County we labored together at Leona. From him I learned much concerning the troubles that hung around the infant association. The enemies of the mission cause pressed them sorely on every hand, and the trial of the brethren in the east, in their efforts to raise the standard of missions, reminded me forcibly of the struggles through which I, with others, passed on this same question, from 1830 till 1834, in the State of Tennessee. For them and the cause they defended I felt an active sympathy; but the pressing need for work on my field, between the Trinity and Brazos, required all my time and talents, and although our organizations in the west were harmonious and active, no preacher could be spared from his field. The death of brother Tryon greatly afflicted us, and we were praying the Lord of the harvest to send more laborers to aid in the cultivation of the vast field. In view of this great necessity, and in answer to our prayers, God sent Elder Jesse Witt to Texas. He was the right man in the right place. With a clear head, full of scriptural learning, and with a large heart inspired fully with the mission spirit, he settled, in 1847, in San Augustine County.

He was by birth Virginian, received a good education, and in early life entered the ministry. He served for some time the General Association of Virginia, as agent for domestic missions. Leaving this field, he accepted an appointment of the Southern Baptist Convention, and came to Texas as a missionary. One year was spent in San Augustine County, and in 1849 he accepted the care of the church at Marshall.

This church was organized in 1847 by Elders John Bryce and G. W. Baines, with twelve members. Of this and the Eight-mile church he served as pastor for several years, with a gradual but steady increase of membership in both bodies. What time he could spare from his pastoral work was spent among the churches and in destitute localities in the surrounding country. He was gifted with much ability in strengthening and encouraging churches; and on that field, at that time, there was great need of such labor.

He was employed as general agent of the Eastern Baptist Convention in November, 1856, which position he filled till June, 1858. He came to Texas at the age of fifty, and spent eleven years in the State under the pressure of infirmity. His feeble health required his resignation in June, and on the twenty-first of November, 1858, he died at his residence in the town of Marshall.

He arrived on the field in good time to aid the brethren in the east who were struggling against the tide of Antinomian and anti-mission principles. As true a missionary as ever waved the banner of the cross, he took high ground, and by the power of speech and patient toil did much to develop the mission spirit in Texas.

In 1849, in the month of May, at the second session of the Baptist State Convention in the city of Houston, I first met him. His personal appearance did not indicate to me the intellectual strength with which I soon found he was possessed. He was a man of medium size, with an open, candid countenance. With an erect body and head, he looked the man whom he addressed in the eye, and yet with his manifest boldness he wore upon his face an affable, gentle and benevolent smile. The kindness and gentleness of his manner drew the heart tenderly towards h him.

In the pulpit, notwithstanding his age and infirmity, he was a tower of strength. Although he spoke with great deliberation, never losing control of his voice, he was heard distinctly by large congregations for hours at a time. Some of his sermons were quite lengthy; but he was one of the few men who could continue to instruct and at the same time awaken deep emotions through a long discourse.

I can never forget his sermon before the Baptist State Convention at one of its sessions at Independence, and I presume that many who read this humble testimony of his worth will remember it. His text, "Ye cannot serve God and Mammon," was carefully and deliberately analyzed, and as he dwelt, for over two hours, upon the utter folly of an attempt to combine the service of God and the world together, every Christian heart was melted to tenderness and inspired with renewed fidelity to God. But few men can weave around the cross of Christ so much sound, strong doctrine in one sermon as did Jesse Witt. He passed to his reward ripe in years and usefulness.

At the second session held with the Macedonia church, Panola County, the name was changed from "Eastern Missionary" to "Soda Lake Baptist Association," with Elder Lemuel Herrin moderator, and William Davenport clerk. Three ministers were present; Èlders Herrin, Lewis, and J. M. Perry. This is the first notice I have of Elder Perry in Texas. He served as pastor at Providence, Border, and other points in Harrison County, for several years, and removed to Hillsborough in Hill County. I met him first in the Trinity River Association in 1856. He gave evidences of earnestness, zeal, and great usefulness as pastor, while a member of this body. In 1858 he was an active pastor of four prosperous churches in the Trinity River Association, and afterwards settled as pastor in the bounds of Union Association, where he still resides.

At the organization of this association there were four churches, with a membership of seventy-seven. At the second session, in 1848, there were eight churches, with a membership of one hundred and twenty-five. The Enon church, Upshur County, that was received at this meeting, was organized by Elder David Lewis, in 1848, with four members.

In 1850, at the fourth session, with Elder Jesse Witt as moderator, the report of the executive committee shows that the churches were engaged in active missionary operations. Elder J. M. Griffin served with success the previous year for his whole time, at a salary of four hundred dollars. This was an earnest of the grand results that followed the operations of this body in the years succeeding, embracing a large territory and building up many churches.

In 1850, the name of Elder Obadiah Dodson appears among the list of ministers in this association. The first work performed by him, that I have any information of, was the organization of Harmony church, Rusk County, assisted by Elder Ray, in 1850, with fifteen members.

With his brother I had an intimate acquaintance in Tennessee, in 1826, and labored much with him there up to 1835. He was always an active worker, and in charge of from two to four churches during all the time. When I first knew him he was much opposed to the mission combinations, but in a few years became an active, thorough-going missionary, in theory and practice.

He was a man of some ability and great zeal, mixed with many eccentricities. He appears, in 1851, among the active workers in the Soda Lake Association, as the pastor of five churches.

The little band, as before recorded, in withdrawing from the old Sabine Association was greatly in the minority, but as John the Baptist said, when speaking of Christ, "He must increase, but I must decrease," so the Lord of the mission cause decided that the new organization should find many friends, while the body opposing earnest combinations of churches to spread the gospel was passing away.

The cause of education met with a hearty response whenever plead in this body, and right vigorously did the mission cause go forward at every session. In 1856, nine years after the organization, the four churches were multiplied into thirty-seven, scattered over the counties of Harrison, Upshur, Wood, Rusk, Cass, Panola and Titus; and the seventy-seven Baptists, who formed the organization on strictly mission principles, had increased to sixteen hundred and thirty. The body still prosecuted its chosen work, but the territory was lessened and churches dismissed to form other associations.

Long live the Colorado and Soda Lake Associations, who were born, as it were, twin sisters, nearly a quarter of a century ago, and have survived the temporal and spiritual conflicts that have swept over the country!

CHAPTER XXIII.

TWO ORDEALS: ONE SPIRITUAL, AND ONE CARNAL. — 1847.

BAPTISTS are a peculair people, and when the mission spirit leads them north to occupy new and destitute fields they certainly exhibit the principles, both in faith and practice, that link them on to the early churches planted by the apostles of the Son of God. Holding the belief that the kingdom of Christ established on earth was not temporal, but spiritual; accepting the declaration of Jesus, "My kingdom is not of this world," as conclusive of the fact that only the spiritually-minded are prepared to become members of it, Baptists in all ages have ever contended for a converted membership in the churches, to the exclusion of all others. As error has no fellowship with truth, and as darkness flees from light, so the sinner in his unregenerated state can have no just conception of, nor sympathy with, the spiritual things belonging to the church of God. Here is the line that divides God's people and His enemies. The regenerate are his, while the unregenerate possess carnal minds, which are "enmity with God."

Baptism is the public profession of this inward change and the declaration of belief in the burial and resurrection of Christ. It is an act of positive obedience to the law of Christ. The immersion of a believer in water being clearly taught both by precept and a number of examples, Baptists adopt it and submit to it as an ordinance from heaven. As one mind governed all the New Testament churches, the same qualifications were required of all who asked for membership, and the same act was submitted to in baptism. To suppose otherwise would be to suppose that there was diversity of opinion and contention among the early preachers, as there is in modern times, with reference to the subjects and action of baptism. We have no evidence of disagreement. Perfect harmony prevailed on both the subjects and the action.

As belief in Christ precedes baptism in the cases recorded, this order must be practiced exclusively or not at all. If any infants may be baptized, then may all as well, and in that case believers' baptism would be done away. If any unbelievers may be admitted through the ordinance of baptism into the church, then may many others be admitted as well, and the world is turned into the church. Christ says, "My kingdom is not of this world."

For their rigid adherence to these principles, requiring candidates for membership first to give evidences of regeneration, and then to submit in faith to the burial of their bodies in water, and, rising from the liquid grave, pledge themselves to walk in newness of life, Baptists have been much opposed in Texas, and in every other land where the influences of Catholicism are felt. He who opposes us on these grounds may with propriety re-examine his positions, and determine whether his doctrines are sustained by the Bible, or whether they are traditions descended from Rome.

These certainly are vital questions entering into church polity. "Can two walk together except they be agreed?" The true churches of Christ are patterned after the model church at Jerusalem. If the character of the membership and the manner in which they were initiated be, both or either, condemned by the word of God, then the claim of the organization to be a true church and authorized to administer the ordinances is vitiated. The Lord's Supper is a church ordinance. It was never intended for the world. Those who partake of it in an acceptable manner must themselves be in sympathy with the sufferings of Christ. The world can't enter into this, and therefore the ordinance was not intended for its observance. As believers are required, in baptism, publicly to avow their faith as the first step in the line of duty, both regeneration and baptism precede the right to approach the table and eat the bread and drink the wine in memory of their dying Saviour.

Christ left these ordinances in the hands of His church when he ascended to glory. As in all well-regulated governments certain official acts can only be performed by officers appointed for the purpose, so, in the church of Christ, he left officers to administer his ordinances. These officers, coming under the hands of the presbytery by order of the body of which they are themselves members, are authorized to administer the ordinances of the church. That everything may be done in order and by proper authority, the body claiming the right to administer ordinances must be composed of proper membership, with the ordinances as they were given, and with officers set apart according to the statutes of the King in Zion.

Many difficulties, as already written, were in the way, both east and west, of perfect organization in Texas. The labor in a new country upon the part of the ministry is very great, and when the gospel has been preached, sinners converted and baptized, and churches formed, the trouble does not cease. In 1846 and 1847 we passed the crisis in the organization of the great Baptist family of Texas.

Campbellism and the anti-mission element gave us the principal troubles up to this time. But now we passed the dangers hanging round the question of *Alien Immersion.*

The sweeping tide of immigration brought Baptists to Texas equally as fast as they could be harmonized, and in some instances a little faster. Brethren in the ministry, many of them, were so anxious to secure the co-operation of all, that a great temptation was offered to receive all adult persons baptized upon a profession of their faith, without questioning the qualifications of the administrator of the ordinance of baptism. Persons coming from Tennessee, Kentucky, Virginia, and other States, where they had been brought up under Campbellite and Pedobaptist teachings, were in some instances thrown among our Baptist communities in Texas, and after personal association with them, and an examination of their leading doctrines, loved the Baptists and their principles, both of which they formerly hated, and were willing to become members among them. In all these cases they were willing to give up sprinkling and pouring, but where they had been immersed by Pedobaptists and Campbellite preachers they were, in many instances, unwilling to give up their former baptism as illegal, and submit to the ordinance at the hands of the Baptists. Some of us understood that the recognition of such baptisms as legal involved the legality not only of the ordination of the minister who performed the act, but also the gospel order of the organization that authorized him to baptize. To admit this much sapped at once the foundation of Baptist faith touching a converted membership and immersion of believers, by authority of the churches, at the hands of their legally appointed ministers.

The decision, however, was made in favor of those who advocated a strict construction of the laws of the kingdom of Christ, and since that day the Baptists of Texas have never been troubled with the doctrine of *Alien Immersion*, and I devoutly trust may never be disturbed with it again.

Having been sorely troubled by the agitation of this question, the decision in favor of what I believed then, and yet believe, to be truth, gave my mind much rest, and I now devoutly thank God that during the past quarter of a century my lot has fallen among Baptists strongly bound toegther by a common faith on all the great questions of church polity. Long may this union mark our progress, as we increase in number.

1847 was a year in which our ministry throughout the State was very active, and as the brethren in the churches co-operated earnestly with their preachers, the cause was greatly prospered. Two associations and many churches, as has been seen, were organized in this year, and the divine favor rested upon the land.

This was my second year as missionary, under appointment of the Board of the Southern Baptist Convention. My appointments were regularly filled, on the field now occupied by the Trinity River Association, and my labor, under great encouragements, was excessive, resulting in the fall with failing health and strength.

Returning home from one of my monthly tours under the burning sun of

August, I found myself greatly exhausted in consequence of a ride of one hundred miles, from Providence church, Navarro County, north of Chambers' creek. After a little rest, I mounted my horse, with gun in hand, with a view first, to look after the farm, and in the second place, if possible, to get a deer or turkey, as fresh meat was called for. The farm was in the Brazos bottom, and at this season of the year the weeds were about four feet high. Passing round the field, I watched every motion of the weeds, expecting every moment to see a deer or turkey. Presently my attention was called to my right, and about thirty steps from my path my eyes flashed upon the head of an old bear, standing on her hind feet, and looking straight at me. My horse was wild, and I dared not shoot from my position in the saddle. Leaping to the ground as quickly as possible, my rifle was levelled, and the mark at which I aimed was just behind the left shoulder of the animal, that was "black as the tents of Kedar." Just as I was in the act of touching the trigger, my game disappeared behind the weeds.

The weeds in a moment shook nearer by, and out ran two young bears, not more than ten feet from me, both of which went up a hackberry tree, near at hand. Resting among the limbs, they turned their anxious eyes on me, that made them look as wild as monkeys in Central America. The old bear was gone, and very deliberately I tied my horse, and with a smile on my face, with none but the bears and the God of the universe in hearing of me, said, "Well, I am good for you certain." My gun was soon to my face again. But just as I was placing my finger on the trigger the second time, the case of David Crockett flashed into my mind, and his unfortunate condition, when he shot the cub and the old bear came upon him, with his gun empty. With this distinguished bear-hunter I had gone in the bearchase in Tennessee. Well for me it was that I thought of my friend David just at this moment; for I had no knife nor dogs to help me out of a similar difficulty. The mind, swifter than electricity, called up in connection with Crockett, the case of David the King of Israel, who when a lad slew a lion and a bear; but as I did not feel willing without a weapon, except the empty gun, to take a bear by the beard, I lowered the gun, and unsprung the triggers.

Just as I lowered my gun a snarl, a growl, some kind of noise that is a nondescript to this day, fell upon my ears, only a little distance off. The old bear, without doubt, was after me. The weeds cracked and shook, and soon she stood on her hind feet, and, in an angry, threatening attitude, walked erect towards me, turning first one side and then the other. Her hair was all turned the wrong way as her ears laid back, presenting a frightful appearance.

Life was now pending in the coming contest. Either Z. N. Morrell or that bear had to die, and my only chance was to make a good shot. To be torn in pieces by a wild beast was a fearful thing; and I shrank with terror at the thought. The bear was not more than forty feet from me and steadily but slowly advancing. The days of flint and steel locks had passed, and remembering that my caps were too small and sometimes failed to fire, I kept my eye on the bear, and let my hammer down firmly on the cap, pressing it well upon the tube. By this time I had what old Texans called "the buck ague."

My nerves were now all unstrung, and for my life I could not hold my gun steady, as I pointed it towards the bear. I remembered on the spot that I had faced the cannon at Hondu and been in danger before, but I never felt as I did when facing that bear. I gripped the gun; but the tighter I gripped the worse I trembled. The bear was now less than twenty feet distant, walking on its hind legs straight to me.

By waving the gun up and down I finally succeeded in getting the range of the body, and not until the animal was within less than ten feet of me did I succeed in getting a steady aim that rendered it safe to shoot. The bear was in the very act of springing on me when the gun fired. At the crack of my rifle the bear, which had received a shot near the heart, ran away, and my trembling ceased, knowing from the manner in which she ran off that the shot was fatal.

I reloaded as quick as possible, and standing on the same spot, shot twice more and brought down the two cubs. The gun was reloaded, and as I started in search of the old bear, which I was confident lay dead among the weeds, a third cub ran up a tree near by, and when my gun fired again it shared the same fate as the rest. The old one was found dead within less than one hundred yards. After gathering the cubs in a pile I sat down to rest from the intense excitement. No battle that ever I witnessed made me tremble as I did on this occasion. In the midst of great emergencies the best policy sometimes requires that we "make haste slowly." Deep impressions were made upon my mind by this incident, that have been of service to me since.

My labors in the fall of 1847 were very arduous. Two long trips were made; one to attend the session of Union Association in the city of Houston; another to Lavacca County, to aid in the organization of the Colorado Association. My health was rapidly failing, and my resignation as missionary of the Southern Board was tendered, to take effect on the first of January, 1848.

CHAPTER XXIV.

TWO ASSOCIATIONS. — 1848

THE events recorded in 1848 show that progress was written upon the banner the Baptist carried in those days. Far up the Trinity and Brazos Rivers, in the direction of the Indian country, they pressed their way, and God smiled upon their work. Elder N. T. Byars, as true and as laborious a pioneer preacher as ever wielded the Jerusalem blade, vigorously prosecuted his work along the western banks of the Trinity, as far as he could find a family to listen to the story of the cross. East of the Trinity and above the territory occupied by the Soda Lake Association, Elders Pickett, Briscoe, Piland and others were busily at work, laying the foundations of the Red River Baptist Association. As the number of associations was rapidly multiplying, and as the Baptists of Texas saw the clearest indications that "the Captain of their salvation" and their glorified Leader intended that they should occupy the vast field and combat error and sin at the very outposts of civilization, north and west, they were now busily engaged in the centre of their operations laying the foundation of the Baptist State Convention.

On Friday before the third Sunday in July, 1848, messengers from six churches — Leona, Society Hill, Springfield, Union Hill, Corsicana and Providence — met in convention with the Providence church, Navarro County, to consider the propriety of organizing a new association. The only ministers present who took part in the deliberations were Elders N .T. Byars and Z. N. Morrell. Elder Henry Hurley, an anti-missionary, whom I had known in Tennessee, was present and preached, but took no part in the organization.

The Providence church with which we met, as has already been recorded, was organized by the writer, in 1846, with fourteen members.

Elder Z. N. Morrell, after a sermon, was called to the chair, and Alexander Patrick was requested to act as secretary. The little body, representing only six churches, with a membership of about forty Baptists, meant work, in the heart of a vast and destitute field.

On Saturday evening, at four o'clock, the committee appointed the previous day reported Articles of Faith and a Constitution, both of which were adopted.

The first article in the Constitution declared that "This association shall be known by the name of Trinity River Baptist Association." The association was immediately organized by electing the same moderator and secretary that acted for the convention, with the addition of Elder N. T. Byars as corresponding secretary, and brother C. B. Roberts treasurer.

The following note appears at the close of the minutes of the first session: "The reasons we assign for organizing so small a body are as follows: — The boundary from north to south is one hundred and fifty miles, and some of these churches could not be represented in the old association (Union) in consequence of the distance; and, moreover, there are four churches still north that have not as yet united. It is hoped that this will attract the attention of our brethren in the ministry, and that they will visit this region of country."

No discordant note was sounded in this organization, and no dissatisfaction with previous organizations caused the formation of this body. Peace and activity marked the first session, and gave a new impetus to the cause in that territory.

The third session was held with the church at Union Hill, in Dallas County. The churches at this meeting, in 1850, numbered ten, with a membership of one hundred and seventy-two. The territory now reached from Bell to Dallas County. Elder A. Ledbetter was chosen moderator. He served as moderator of this body for three successive sessions, and during these same years served as missionary of the Batpst State Convention.

Elder Ledbetter moved from Tennessee to Texas in 1848, and settled in Dallas County. Although a man of limited education, he possessed many of the elements of usefulness, yet lacked that stability of character and stern integrity so essential for a preacher. While he occupied the chair as moderator of the association no special indications were given of unsoundness in the faith of the gospel. In the spring of 1854, at Springfield, in my presence, and in presence of the members of the Mission Board, he preached a sermon strongly tinctured with apostasy. Fearing I might be mistaken, I made no charges against him, nor did the brethren with me. Soon after this, in the presence of brother Byars, he advocated the doctrine in a sermon, boldly. Of this I was immediately informed by letter.

At the seventh session, in 1854, held with the church at Centreville, at which time Elder Ledbetter was under appointment to preach the missionary sermon on Sunday, I determined, after consultation with other brethren, to make an issue. Accordingly on Saturday morning I sought an interview with him, in company with Elders G. W. Baines, J. W. D. Creath, and four other preachers. In the presence of seven ministers the accusation was brought, on the ground of heresy. He hesitated for a moment, but finally admitted that he believed the doctrine of apostasy, and that he had always believed it, and further indicated a desire to debate the question in presence of the association.

Had he embraced the doctrine at a recent date, there might have been grounds for leniency and charity; but when he announced his position as a believer in apostasy even when he joined the association, after having given consent to the article on that subject for several years, one of the members of this self-appointed committee charged him openly with hypocrisy and lying. Elders Baines and Creath each gave him the benefit of some plain Baptist talk, and his right to preach the missionary sermon on the following day was challenged.

The same evening he publicly withdrew from the association, and, after the body adjourned, left for his home, sowing the seeds of discord as he went.

We had previously been troubled with Elder C. C. Owens, who claimed to be a Baptist, and yet divided a church, leading a part off into Campbellism. There was another preacher travelling among us by the name of Stroud, bearing credentials as a Baptist preacher, and teaching the doctrine of Universalism. A third, by the name of Jeffreys, an impostor, had thrown us into some confusion. These all had been dealth with promptly. When Elder Ledbetter appeared as the fourth case, rapidly succeeding the others, no wonder that some of us, possessed of a nervous temperament, should grow a little impatient and make a vigorous prosecution.

Among the old-time Baptists, with whom I labored much in my early ministry, and who preached so strongly the doctrine of salvation by grace, a man who intimated that works had anything to do with salvation was called "linsey-woolsey." This expression was applied in view of the fact that the priests, in ancient times, were not allowed to wear garments part of wool and part of linen. They were required to be either all wool or all linen.

Men who would sacrifice principle for the sake of popularity were sometimes called "tin-headed." As tin is a very soft metal, susceptible of being easily dented without breaking the surface, so a preacher, whose head was soft enough to be marked by the hammer of popularity, soon had this appellation tacked on to his name.

In the history of several of our Texas Baptist associations, while we have been held in check for a time by the honest, hard-headed, "iron-jacket" and anti-mission fraternity, we have been wonderfully cursed with these "linsey-woolsey, tin-headed" Baptists, who, in the language of Paul, have been "those who seemed to be somewhat, — whatsoever they were it maketh no matter." Elder Ledbetter's was the last case of heresy that gave us trouble in the Trinity River Association.

At the third session, Elder H. P. Mays made his first appearance among the Baptists in a general meeting. Here was one of the brightest stars that ever shone in the Baptist galaxy in Texas. The star remained only a little while, but was bright while it lasted.

Elder Mays was, by birth and education, a Kentuckian. He came to Texas in December, 1849, and settled at Corsicana. He died December 4, 1851. In 1850 and 1851 his name appears among the list of ministers present at the association.

I first met him at Springfield, the year after his arrival, and was at once deeply impressed with his manner of address and great personal piety. He was truly an educated, Christian gentleman. In person, he was tall and commanding. He was sound in doctrine. The social element entered largely into his composition. While deeply earnest in his work as a preacher, his chief excellence consisted in his ability to lead the minds of his audience, melted into tenderness, right up to the cross of Christ. His short, earnest life of two years in Texas left a deep impression upon the community, the church, and association of which he was a member.

The minutes of the seventh session show that there were twenty-six churches, with a membership of five hundred and fifty-seven. At the eighth session, held with Little River church, in 1855, the first meeting after the difficulty with Ledbetter, thirteen newly-organized churches sent in petitionary letters and were recieved into the body. This was in a large measure due to the labors of Elder John Clabaugh, who was appointed by the Baptist State Convention as financial agent. Turning aside from this work, he penetrated the destitute corners of McLennan and the adjoining counties, and, instead of collecting money to send other missionaries, he preached, baptized the people, and organized them into churches himself. Some complaint was made against him for leaving the special work assigned him, but when the result of the mission was reported, all agreed that he had done his duty. In 1855, God wrought wonders in this territory. Elders Eaves, Mason and McLain were all active in sowing the seeds of truth, and their work was wonderfully blessed.

The thirteen churches received that year had a membership of two hundred and seventy-seven. At the organization, in 1848, our whole strength was forty; but now, in 1855, we received an added strength, at one meeting, of about seven times our number at first.

The name of Elder S. G. O'Brien appears at this time among the active workers of the body. But few drones were in our hive in those days, in the bounds of the old Trinity River Association. Ours were worker-bees. God have mercy upon a lazy, stupid Baptist, especially if found in the ministry. He may be fit for something, but I can't tell for what.

Elder O'Brien came to Texas in 1852, and took position in Baylor University as Professor of Mathematics. In the fall of 1853, I received information that he was willing to resign his position in the school, where he received a salary of twelve hundred dollars, and accept a position as missionary or pastor, if one half this amount could be secured for him. I had never seen him; but this made a deep impression upon my mind, as we were greatly in need of ministers among our feeble churches, and I wrote to him and urged him to visit us. He came, and, in the spring of 1854, settled at Waco, and for six years served that church in connection with others in the surrounding country.

The Waco church was organized by Elder N. T. Byars, on the thirty-first day of May, 1851, with four members. Elder Byars served as pastor till the arrival of Elder O'Brien, at which time the church numbered nineteen members. Under his ministry this church steadily increased in number and efficiency. In 1860 it numbered one hundred and forty-four.

Progress was written upon the banner he carried, and, being full of the mission spirit, his influence added much to the missionary operations of Trinity River Association. He was a man of education, and believed, as a Christian, that it was our duty to train the mind for effective service in the Lord's vineyard. He was the first president of the Trinity River High School, located at Waco in 1856, — now known as Waco University, — and may justly, according to my judgment, be regarded the leading spirit among its founders.

He subsequently served as pastor at Huntsville, filled for a time the presidency of Bosqueville Academy, and in September, 1867, in his forty-seventh year, while serving the churches at Port Sullivan, Cameron and Little River, died.

His appearance, in public or private circles, commanded respect. Something over meidum height, he weighed about one hundred and eighty pounds. His manners were simple, natural, and earnest. He possessed a free, open countenance, and was so perfectly conscious of his own honesty of purpose that he did not suspect others. If this be a fault, he was guilty. Let anybody who ever blamed him for it remember that the purity of the man's own motives was at the bottom of it.

He had a clear head, and was as sound in the faith as any preacher who has ever opposed the systems of error propagated in Texas. He was a student, and a man of God, and in the pulpit made no failures.

At the eighth session, a resolution, declaring that it was the sense of the body to organize two high schools, one for males and the other for females, was adopted. In 1856, the association located the Trinity River Associational Baptist High Male School at Waco, and the Female High School at Hillsborough. Concerning the school at Hillsborough I cannot write, as there are no statistics before me. Nor am I able to state definitely some of the changes that have taken place in the school at Waco.

Elder Rufus C. Burleson succeeded Elder O'Brien in the presidency at Waco in 1861, carrying with him an able corps of teachers. The school has subsequently been known as Waco University, and has attained a degree of prosperity second to no institution of learning in the State. In the heart of a rich country, and in the midst of an intelligent and enterprising community, and with such leaders as Elders R. C. and R. B. Burleson, nothing short of success could have been expected.

At this institution, under the charge of President Burleson for the past ten years, a large number of students, of both sexes, have received tuituion. Much has been done and much is still being done for the education of the rising ministry of Texas.

The Trinity River Association has done much for the cause of missions, and a large number of churches owe their origin, under God, to the labors of her missionaries.

The same meeting that laid the foundation of the school at Waco appointed a committee to collect funds for the support of her superannuated ministers. This committee continued its operations, whenever cases came up requiring assistance, and in process of time gave much relief.

The little band organized in July, 1848, with two ministers, six churches, and forty Baptists, passed at its first meeting the following resolution: "That this association set apart Friday before the third Lord's day in May, as a day of fasting and prayer to Almighty God, that he would grant a greater effusion of the Holy Spirit to the little churches composing this body, and that he would more abundantly bless the few laborers we have among us; and that he would send forth more laborers into this part of his moral vineyard,"

In answer to prayer, after much earnest labor in the cause of Christ, the minutes of the tenth session, in 1857, show that there were, in this body, twenty ordained and eight licentiate ministers, and thirty-four churches, with a membership of about twleve hundred. A large number of churches have gone from this body, at different times, to aid in the formation of other associations, but still a large and rapidly increasing body lives, bearing the old name, and prosecuting its mission work. About four hundred baptisms, performed by the hands of her ministers, were reported at her last session in 1871.

Elders W. H. Parks, G. W. Green, and T. S. Allen were active and laborious preachers in the midst of these revival influences.

Elder Parks was educated at Baylor University. His field of labor was for some time in Northern Texas. He subsequently moved to Freestone County, and under his ministry the church at Fairfield and others in the vicinity have been greatly strengthened. He is full of zeal in the cause of his Master, and, being in possession of an active mind, promises extensive usefulness.

With Elder Green I have been intimately associated much of the time since he entered the ministry. He was ordained in 1863, and was for some years a missionary in the Trinity River Association. He has since served as pastor at Bedais, Oakland, and other points in Grimes, Leon and Brazos Counties. Whether as missionary or pastor, his ministry has always been of that effective character that rouses the people to action touching the great question of personal salvation. But few men, in so short a time, have baptized so many people in Texas as Elder Green. He leans upon the arm of God in earnest faith, and the Spirit of God applies the truths he utters.

Elder T. S. Allen, the old soldier who fought on so many hard-contested spiritual battle-fields in Missouri, was of that class of Missouri preachers who, during the late war, found it necessary to seek another clime. With his loins grided and staff in hand, he journeyed to Texas, and on the territory of the Trinity River Association displayed his colors "as a good soldeir of Jesus Christ."

I met him soon after his exile, while his family was yet behind, and he was in the deepest trouble. He was not forsaken by the Master whom he served, and the loved ones soon joined him on his new field. With a vigorous constitution, and a soul deeply stirred for the salvation of sinners, his motto, Sunday, Monday, every day, is, "Work, work for Christ." Some preachers, like many Christians, work at times as though they expected the return of the Master to-morrow, and then again as though they thought it doubtful whether or not he ever would come. Not so with Elder Allen. As in Missouri so in Texas he goes along the creeks, into the lanes and by-ways and hedges, and, as missionary and colporteur, preaches to many or few, as he can gather them together, leaving by the way a Bible, a religious book, or a tract, always confident that "labor is not in vain in the Lord." By him large numbers have been baptized. Oh that the Lord would send more such workers to Texas!

Three months after the organization of the Trinity River, a number of ministers and brethren met at Honey Grove, in Fannin County, in the extreme north-eastern part of the State, and near Red River, for the purpose of forming an association. Eight churches — Clarksville, Shiloh, Salem, South Sulphur, Liberty, Bethel, New Salem, and Honey Grove — were represented by their messengers in this convention. These churches were located in the counties of Bowie, Red River, Lamar, Titus, Fannin and Hopkins. Elder Benjamin Clark was the first moderator.

This venerable brother came to Texas only a short time before. He was the first missionary sent to Missouri by the Board of American Baptist Missions. On that field he baptized a large number of people and organized many churches. He brought with him to Texas the minutes of eighteen sessions of a Missouri Baptist Association, which I was permitted to examine, and which exhibited a vast amount of labor performed by this venerable missionary. He spent several years in Arkansas, and arriving in Texas spent some time among the churches and brethren that formed the Red River Association.

Elder Clark came into the bounds of the Trinity River Association about 1852, and there, while the sun of life was passing behind the western hills, he reflected back much of the light he had borrowed from Chirst. He was about seventy years of age when he came among us, but, being provided by the brethren with a home, he did much active service as a missionary previous to his death. The infirmities and necessities of this good man caused the formation of the executive committee alluded to before, whose duty it was to provide for the maintenance of superannuated ministers. He remained on this field several years, and passed to his reward above.

The second session of the Red River Association met at Clarksville, Red River County, on the twelfth of October, 1849. A petitonary letter was received from the Liberty church, Titus County, and the statistics show that the total membership of the churches was one hundred and seventy-five, leaving out the membership of the South Sulphur church, Hopkins County, which was not represented. Elder M. Piland was moderator. Of this brother and his work I cannot write definitely, except that his name appears for some time as pastor of Shiloh church, Lamar County, and as moderator of the association at its second and third sessions. The association devoted itself energetically to the mission work, and in connection with missionary operations the name of Elder W. M. Pickett is pre-eminent.

Elder Pickett came to Texas in 1844, and was a leading spirit among those who labored in the work of organization along the banks of Red River. He was present at the formation of the association, and appears on the record of 1849 as pastor of five of the churches when there were only eight churches represented. He devoted all the time he could spare from the churches to destitute fields, and in 1853 was appointed missionary for his whole time, to visit destitute regions, organize churches, and to do all he could in "procuring pastors for destitute churches." His salary was fixed at four hundred dollars. Long and faithfully did he serve in this capacity, as is seen from his reports, and his work was greatly blessed.

The name of Elder John Briscoe appears among the Baptist workers in north-eastern Texas, and continues to shine as a bright star in the history of that people. He was by birth a Tennesseean; moved to Texas in 1846, and settled in Hunt County. He was at this time a licensed preacher and giving signs of great usefulness. The Liberty church, Fannin County, called him to ordination. The presbytery was composed of Elders Pickett, Watson, Smith, and Piland. He took an active part in the organization of the Red River Association, and in 1850 appears as pastor of Liberty church, Fannin County, and also of the Salem church, Lamar County. His soul was deeply moved whenever the cause of missions demanded his attention, and, being a man of strong mind and much decision, he was ordained of God to be a leader among the scattered Baptists in that territory.

In 1852 he was moderator of the body, and the same year employed to ride as missionary one half his time. After the formation of the Sister Grove Association he served as pastor and missionary among the churches of that body. His report to the body, in 1854, shows that he had witnessed the conversion of one hundred and twenty persons, and that he had baptized eighty-five, within the space of three months and ten days. During this time he travelled six hundred and thrity miles and delivered eighty sermons and exhortations. The date of his death I cannot give; but while he rests from his labors, his works follow him.

Under the labors of these earnest men the sixth annual session of the body, held with the Concord church, Red River County, in 1854, was composed of messengers from twenty-eight churches, extending over a territory of seven counties. After this a large number of the churches went off to form other associations, and in 1858 only eleven churches were represented, with Elder D. B. Morrill as moderator. The territory was at that time limited to three counties: Lamar, Red River, and Bowie. Later than this I can give no statements, for want of facts and dates.

CHAPTER XXV.

STATE CONVENTION. — 1848

IN all the great enterprises of life, whether civil or religious, a necessity is felt, when there are small organizations in a territory, for a general organization to harmonize and concentrate the efforts of the whole body. When the noise of war is heard in the land, companies at once are banded together for the public defence. After this, battalions, regiments, brigades and divisions spring into existence. All these under the same laws, and under the same great leader, present a strong, united front to the common enemy.

Ten years previous to the organization of the Baptist State Convention of Texas, churches had been formed. Eight years previous to the formation of this grand body, an association of churches was formed; and now that a vast field had been traversed by missionaries, a great many churches organized, and several associations, a general organization was called for, in which messengers from churches and associations might meet every year, and by conference and co-operation sweep over the whole State, and following close upon the heels of the Indian and buffalo, plant the standard of the cross wherever the smoke of the white man's cabin rose toward the heavens. Many a warm and generous Baptist heart beat true to this sentiment in the territory of Texas, in 1848. The few scattered preachers, sustained in their efforts by a faithful brotherhood in the churches, said, "We will attempt great things and expect great things for God." Truly God had done great things for us, and we were glad. Far more had been accomplished than I, and the little bands that met me in 1836, anticipated could have been done in the space of time.

An incident in the life of Elder Jesse Mercer, of Georgia, throws some light upon the origin and progress of Texas missions. Some time previous to 1840, the date not known, this great and good man deposited twenty-five hundred dollars with the Home Mission Society, New York, to be used for the support of missionaries in Texas. Some of his friends protested against the expenditure of money on such a field. He was informed that Texas was at that time infested with thieves, murderers, and scoundrels, who were refugees from justice. Elder Mercer was a man who thought and acted for himself, in view of his personal accountability to God, and replied, "You had better not tell me any more about such characters in Texas, or I'll be compelled to double the amount and set apart five thousand dollars." He stated his conviction, that, as Texas had a fertile soil and a genial clime, it would attract the attention of a large number of good people. Christ had saved a thief on the cross, and if some of those in Texas were great sinners, Christ was a great Saviour, and they needed the gospel.

With that money donated by Elder Mercer, Elders Tryon, Huckins and Taliaferro were sent and supported in Texas. The first person baptized by brother Huckins, on his arrival in Galveston, was a relative of Jesse Mercer. I allude to sister Borden. Old brother Eli Mercer, another near relative, and the father of sister Borden, came fifty miles from his residence in Egypt, on the Colorado, to Mount Gilead church, in Washington County, to be baptized by brother Tryon. He continued his membership with that church for some time, and regularly rode the fifty miles, and filled his place in the monthly conference meetings.

The prayer and alms of Jesse Mercer went up as a memorial before God, as in the case of Cornelius, and God bestowed salvation in Texas upon some of the members of his own house. Some of the results that followed the labors of Tryon, Huckins and Taliaferro have already been recorded.

At the eighth annual session of the Union Association, held with the Baptist church in Houston, in October, 1847, the following resolution was passed: "*Resolved*, That this association appoint a Central Committee of Correspondence, composed of Elders Graves, Garrett, Ellis, Chandler, Tryon, Creath, and brethren Haynes and J. G. Thomas, whose duty it shall be to receive from the corresponding secretary the information that he may obtain, and in the event that a majority of the churches so corresponded with shall be in favor of a convention, then it shall be the duty of the central committee to appoint a time and place of meeting."

The churches responded in favor of the proposed convention, and on the eighth day of September, 1848, messengers assembled with the church at Anderson, in Grimes County. As is customary at the meeting of all such bodies, an introductory sermon was expected. In view of this fact, the committee that called the meeting appointed Elder Henry L. Graves, one of their number, to preach the sermon, and Elder Noah Hill, in case of failure. These brethren were duly notified of the appointment, and were both on the ground in good health, and as I thought certainly without excuse.

Elder Graves was at that time president of Baylor University, and we all thought the appointment a judicious one, and were anxious to hear him preach the opening sermon. To our great astonishment he declined to do it, without any plausible excuse. This was evidently, to take it as a whole, the most learned body of men that had ever been assembled in Texas up to that time, and the impression of course fixed itself upon our minds that President Graves was afraid he might make a failure and thus sacrifice some of his reputation. Elder Hill refused to preach as alternate, and, under the circumstances, we could not blame him. A number of the talented brethren were urged to preach, and my conviction was that they were the most tender-footed set of Baptist preachers that ever I had seen assembled.

My mortification was intense, as the large audience waited for over half an hour for the services to commence.

I had been used on other occasions, as the reader will remember, as an iron wedge, and sometimes driven into very unenviable situations. Although a little sore at the remembrance of it, when the committee approached me I consented, for the sake of our cause, to preach. Once at least in life my position was easier than that of my brethren who were my intellectual superiors, — Z. N. Morrell had no reputation to lose.

The text used was from Isaiah ix. 7: "Of the increase of his government and peace there shall be no end." After the sermon, the meeting was called to order by Elder R. E. B. Baylor, after the usual preliminaries, notwithstanding the failure to preach the sermon. Elder Henry L. Graves was chosen as the first president of the Baptist State Convention of Texas.

Elder Graves was a part of that valuable cargo of preachers landed at Galveston in the fall of 1846. He enjoyed the benefits of both literary and theological training, and graduated in both departments. As Baylor University, over which he presided as its first president, was founded with the view of disseminating knowledge among the masses, but especially for the education of the rising ministry, he possessed those qualifications and advantages that fitted him well for the position he assumed. After filling that station for several years, he took charge of the college at Fairfield, and for a number of years labored successfully in the education of young ladies. He now fills the presidency of Baylor Female College. His scholarship, so far as I know, remains unquestioned during all these years of patient toil as an educator in Texas. His qualifications entitled him to the position, in the estimation of his brethren, of a refined and educated Christian gentleman.

As a presiding officer he excels. Calm, dignified and courteous, he commands the confidence and respect of the body, and makes perhaps as few mistakes in his rulings as any man who has ever presided, in Texas, over deliberative bodies. In the midst of the most animated discussions he never loses sight of the question, and shows himself master of the situation by maintaining perfect self-possession. He has acted as president of the convention at most of its sessions since.

An incident occurred during his early pastorate at Independence, clearly illustrating the deliberation and decision of his character. In the midst of the first great revival among that people he conducted a lady into the beautiful stream, and, after administering the ordinance of baptism in the most graceful and imposing manner, observed a very poisonous snake lying upon the lady's robe. I was standing near by, and was just in the act of speaking to him, when he suddenly seized the snake in his hand and threw it to the opposite bank. Elder R. E. B. Baylor, standing at my side, spoke with animation and emphasis, "Why, sir, the apostle Paul could have done no more!" The lady knew nothing of the danger till informed of it afterwards.

The second article of the constitution adopted by the convention was as follows: "The objects of the convention shall be missionary and educational, the promotion of harmony of feeling and concert of action in our denomination, and the organization of a system of operative measures to promote the interest generally of the Redeemer's kingdom within the State."

"An old landmark" among Baptists in all ages, touching the sovereignty and independence of the churches, was asserted in the twelfth article. Churches create associations and conventions, and, as the creator has the right to control the creature, in keeping with this fundamental idea, the convention disclaims any and all right to dictate to the churches.

The article reads as follows: "The convention shall never possess a single attribute of power or authority over any church or association. It absolutely and forever disclaims any right of this kind, hereby avowing that cardinal principle that every church is sovereign and independent." They who recognize the authority of popes, bishops, synods and conferences, will doubtless inquire, after reading this article, the necessity for this organization upon such principles. We reply, that while every church recognizes and should recognize "Jesus only" as its perfect lawgiver and chief shepherd, a general organization was necessary to secure harmonious and effective action upon the part of individuals and churches. Here was a body of influential, earnest men organized, upon scriptural principles, to promote the common causes of education and missions, and their individual and combined influence has been telling upon the destinies of this great empire State during the twenty-three years that have passed.

This was the first general meeting in which Elder Rufus C. Burleson made his appearance. When the tall, slender form stood on the floor of the convention for the first time, seeing everybody in sight at one glance with those black, piercing eyes, that rest beneath a manly brow, and, pointing with his long, bony fingers in the direction he wished the thought to travel, parted the lips of an orator, and spoke sweetly and tenderly the name of Jesus, the stranger involuntarily asked his neighbor, "Who is that?"

He came to Texas in the month of February, 1848, under appointment of the Mission Board of the Southern Baptist Convention, and settled as pastor of the church in Houston. As the successor of the beloved Tryon, God blessed his work with much success for over three years in the city of Houston. In 1847, at the time brother Tryon died, the church according to the minutes of Union

Association, numbered sixty-nine; and in 1851, under his ministry, it had increased to one hundred and forty.

As a pastor, he was among the first, and was largely endowed with the elements of enlarged success in this department of Christian labor.

Fired with a growing desire to have the coming preachers in Texas thoroughly educated, at the call of the Board of Trustees of Baylor University he resigned his pastorate in Houston, and took the position at the head of the institution as an educator. He remained as president of Baylor for ten years, commencing his duties there in 1851. The same qualities that made him successful as a pastor gathered about him the affections of the youth of the country, and inspired their parents with confidence in his ability as a teacher. Resigning his position at Baylor in 1861, he was immediately elected to the presidency of Waco University, which position he now occupies, and has, without interruption, for ten years. His ability to build up is known as far as the man is known. His brother, Elder R. B. Burleson, has had much to do with the success that has crowned his labors in the cause of education. The combined influence of these brothers has fired many a youthful mind to toil in search of knowledge, and impressed many a heart with the importance of seeking Jesus.

They have educated, either in whole or in part, about fifteen hundred young men, and over four hundred young ladies; and the struggle of the past twenty years in this work has in no wise abated their energy.

As a preacher, Elder R. C. Burleson has wielded an influence over the masses in Texas second to no man who has occupied the pulpit among us. Towards the close of his collegiate course, whilst his associates in the theological department were consecrating their lives to various fields, some of them to foreign countries, R. C. Burleson wrote, "This day I consecrate my life to Texas." This was a noble purpose and a high resolve, and during these twenty-three years that I have known him in Texas no disposition to falter has ever been manifested.

At the second session of the body, in 1849, held with the church at Houston, Elder J. W. D. Creath president, the question of missions was considered with much interest. The Board of Directors was instructed to employ, as soon as practicable, two missionaries, one in the eastern and the other in the western congressional district, with instructions to organize the scattered Baptists, and collect money for mission purposes. Of course, at that time, a vast destitution was before us, and right earnestly did the brethren address themselves to the work of supplying it. But little being accomplished previous to the third session, in 1850, a report strongly recommending a general agency was adopted, and soon the name of Elder J. W. D. Creath appeared as financial agent of the convention. East, west, north and south, over the vast territory, this indefatigable worker travelled, and greatly stirred up the mission-loving Baptists throughout the State. Every subsequent session of the body showed large sums of money collected and expended on this mission work. The names of missionaries engaged under the patronage of the Board, and the work they have done, would require a volume of itself. Perhaps no agency has accomplished so much in developing the Baptist cause in Texas as the mission enterprise of the Baptist State Convention from its organization to the present time.

Elder J. W. D. Creath has been the leading spirit among us in this great work. During most of the time since the adoption of the agency system he has been, and still is, the travelling financial agent of the convention. He was by birth and education a Virginian. Following the example of his father, a Baptist preacher, he and several of his brothers entered the Christian ministry, and, as has already been noticed, he came to Texas in 1846. He first settled as pastor at Huntsville, and during the interval in his agency served at Cold Springs, in Polk County, and vicinity. He has been eminently sound in doctrine and an earnest defender of the faith. Look where you may among the minutes of associations and general Baptist meetings in Texas, and the name of Elder J. W. D. Creath appears pleading the mission cause. Much might be said of him and his work, but the utter folly of an attempt on my part to tell the people what they do not know concerning this laborious preacher is happily illustrated by the following incident that occurred in a neighboring village: —

A widow lady, who kept a hotel for a number of years on one of our thoroughfares, was approached on several occasions by a local editor with earnest solicitations for an advertisement of her house. She modestly declined at first, but finally, weary of his importunity, informed the editor that her house was known farther than his paper. So in the case before us. Elder Creath is not only known farther than the writer, but has made impressions for good in his Master's work in many a locality which these pages will never reach.

The names Elders Ellege, Fisher, Eaves, Thomas, Clabaugh, Kiefer, and many others, appear as zealous missionaries on different fields, co-operating with the convention and spreading the gospel among the destitute. As a specimen of faithful and effective labor I will give an extract from the report of Elder David Fisher, read before the convention. During the year ending in November, 1855, "he preached two hundred and forty-nine sermons, delivered four hundred and eighty-two exhortations, attended one hundred and one prayer meetings, baptized and witnessed the baptism of two hundred and seventy-six persons; and travelled two thousand three hundred and forty miles." This earnest man of God, who always preached in such a manner as to convince the people that he believed there was a hell to shun, a heaven to gain, and a Saviour to redeem, came to Texas in 1846, and settled as pastor in Washington County. He has sepnt much of his time as a missionary, and many a redeemed one will rise up at the last day to praise God for this instrument of usefulness.

In 1857, the aggregate of missionary labor is given. We extract the following:

"Miles travelled, 26,666; sermons preached, 1,920; exhortations delivered, 1,244; baptisms administered and witnessed, 556." This same year Elder Michael Ross came to Texas, and, upon the resignation of Elder Creath, was appointed financial agent. He was a walking encyclopaedia of scriptural knowledge. He had committed to memory many parts of the Bible, and was wonderfully gifted with ability to expound the Scriptures. His mission spirit knew no bounds, and as he brought the word of God to bear upon Christian duty in this line, large amounts of money flowed into the treasury of the convention. He filled this position with much success until 1860. After his resignation he served as pastor in Falls County, and subsequently settled as pastor at Independence, where he died, in 1865. In his grave lies one of the strong men of Israel.

At the fifth session of the convention, held at Marshall, in 1852, Elder H. L. Graves president, a committee of seven was appointed "to take into consideration the propriety of establishing, at some eligible place in Texas, a Baptist paper." Reports on this subject, urging the importance of a State paper, appear in every minute up to 1855, when "The Texas Baptist" was reported to be in existence, with Elder G. W. Baines as editor. The paper was located at Anderson, and was published under the direction of the Texas Baptist Publication Society. A vast amount of good was accomplished by this denominational organ, furnishing as it did a medium of communication for the brethren and churches from the Sabine to the Rio Grande.

It required only a few years, however, to demonstrate the fact that individual enterprise, backed up by the denomination, is far preferable in the publication of a paper to the plan adopted in publishing the old "Texas Baptist." It is true that the paper went down among the lost fortunes during the late war; it is also true that financial embarrassments hung around the Publication Society from the commencement to the end.

The paper developed a large amount of writing talent in Texas, and gave the editor and others an opportunity of discussing a great many questions of vital interest pertaining to our common cause in Texas. The editor, Elder G. W. Baines, has rendered valuable service, not only through this paper, but also as pastor of several churches, and on many tours through the State.

His name first appears in the minutes of 1850. He has served as pastor at Huntsville, Anderson, and other churches in Grimes and Washington Counties. Being naturally fond of metaphysics, he has frequently been spoken of by the brethren, after his sermons and after debates on the floor of the convention, as "The hair-splitter." In faithfulness he still labors as pastor at Salado, Bell County, and among the churches in that vicinity.

Under the auspices of the Baptist State Convention, Baylor University and Baylor Female College have performed their work during the past quarter of a century. Although the territory was too large for the agencies of one body to be effective in every part of the State; although the Eastern Baptist Convention took charge of the territory east of Trinity River and subsequently merged into the General Association, taking charge of an additional territory in northern Texas, — yet a large field still remains for the old body, whose efforts have been so signally blessed in the past. Frequently during late years, as I meet with the old convention, my mind goes back to 1848, and, viewing the organization in connection with the results, I am made to exclaim from the depths of my soul, "What hath God wrought!"

I will record an incident illustrating the character of a native Texas horse, and also some of the trials in connection with this animal that have befallen a number of Texas preachers who have been compelled to rely upon this species of locomotion.

While travelling still in the bounds of Trinity River Association my horse was crippled, and the only chance to meet my appointments was to ride an untrained mustang horse. The animal had been ridden, but was by no means docile. Accordingly the horse was roped, bridled and saddled, and, to prevent the disposition these Texas horses sometimes manifest, of spring as high from the ground as their strength will allow and then descending to the earth with the back in a bow and the head between the forelegs, I secured a tough dogwood forked stick, and tied the large end to the girth and the end of each fork to the cheeks of the bridle. This preparation made, with my usual equipage I mounted, and rode off to fill appointments for one hundred and fifty miles up the country. All was well as long as the stick remained; but I was exceedingly annoyed by questions from almost every man I met. Besides, it was very troublesome to loose and adjust the stick every time I stopped at noon and night.

Crossing Chambers' Creek, I saw that the horse was weary, and, supposing that its propensity to "pitch" was over I untied the dogwood fork and threw it aside. As I approached the house of my old friend Morrell, a few miles below Dallas, the horse, without any cause that I could discover, commended "pitching," or, as the old Texans sometimes said, "laying fence-worm." Freely would I have given the value of the horse for that dogwood stick, well adjusted. Now the horse sprang, first to the right and then to the left, and then came to the ground, head down and heels up, almost in a perpendicular. This performance continued for nearly a hundred yards. My hat flew off, my umbrella fell, my saddle-bags took wings, and I began to get as limber as possible and look for a good place to fall. Just as this juncture of affairs the horse stopped, very much exhausted, and I did not fall. I was so bruised and shocked that I was immediately thrown into a violent fever, from which I did not soon recover. Thus closed with me the mighty events of 1848.

CHAPTER XXVI.

TWO ASSOCIATIONS. — 1849.

THE rapid development of the agricultural, educational and religious interests of Texas, from 1845 to 1849, was truly encouraging. The tide of immigration swept across the State, increasing greatly the strength of villages and settlements, and stretching far out upon the frontier. Bountiful crops owned the labor of the husbandman, school-houses were springing up in almost every community, and the heralds of the cross, keeping pace with civilization, pushed on their work with abundant success.

Six associations had been formed, and the Baptist State Convention. All these organizations, except one, were alive, and putting forth active and harmonious efforts in the cause of truth. There were at this time about seventy-five churches, and, as near as I can estimate from the minutes of that date, over two thousand Baptists, with rapid accessions to our number constantly occurring, both by baptism and by letter.

The work was being vigorously prosecuted far up the Trinity River, and in October, 1849, representatives from four churches met in convention with the Union church, Dallas County, for the purpose of forming a new association. The churches represented were Rowlett's Creek, Union, Bethel and Lonesome Dove. Four ordained and one licentiate minister were present at the organization. The first article in the constitution reads as follows: "This association shall be known by the name of the *Elm Fork United Baptist Association.*"

From some historical sketches of churches in the minutes of 1857, I will state, that the Rowlett's Creek church was organized by Elders David Myres and Jonathan Phillips, on the twelfth day of February, 1847, with seven members. Elder Myers was the first moderator of this association, and continued to preside over every session until his death, which occurred on the ninth day of March, 1853, in his fifty-seventh year.

With him I had no personal acquaintance, but often had interesting accounts of him and his work through Elder N. T. Byars. He moved with his family to Texas in 1845, and settled in Dallas County. On the tenth day of May, 1846, he organized the Union church, Dallas County, with five members. He was the first

Baptist preacher in that part of the country. He assisted in the organization of several other churches, and under his pastoral care Rowlett's Creek, Union and Bethel churches were greatly blessed. The writer of his obituary, in the minutes of 1854, says, "In his preaching he was plain, easy to be understood, forcible in argument, and pointed in application. He dwelt extensively on the plan of redemption and the love of God." These certainly constitute the qualities of a good minister of Jesus Christ, and this the grand central theme around which the preacher should always rally his thoughts. The churches and the association deeply felt the loss when the voice of Elder David Myres was hushed in death.

The names of Eli Witt, J. A. Freeman and J. Phillips appear at the organization. Several sessions passed before the infant churches were able to make appropriations of money to the mission work, but a true missionary spirit fired the hearts of these pioneer preachers, and at the sixth session, in 1854, held with the Bethel church, Elder J. M. Myres, moderator, the report of the committee on Home Missions shows that Elders Freeman and Myres had performed a considerable amount of missionary labor, with good success. At this session of the body there were eleven churches, with four hundred and eighty-three members. The names of Elders Eli Witt, N. T. Byars, G. W. Butler and H. E. Calahan appear as active missionaries at subsequent meetings, and following the work of these and a few earnest pastors, the association, in 1861, had eighteen churches, with about seven hundred and fifty members. The territory of the Elm Fork Association, as defined in the minutes of 1856, was "all of Dallas County north of Trinity River, all of Denton County east of Elm Fork, together with the counties of Collin and Kaufman."

Elder J. M. Myres was moderator of the association in 1854, and at several meetings afterwards. He is the son of the old pioneer preacher, and came to Texas the same year his father came. He was licensed to preach in 1849, by the Union church, and was ordained the same year by Elders J. A. Freeman, David Myres and Eli Witt. He has rendered much valuable service both as missionary and pastor in the bounds of this association, and still lives and labors in Dallas County, on the same field where his father did so much valuable work in organizing and building up churches, and from which he was called to serve his Master in a better land.

In 1853 Elder J. C. Portman came to Texas from Kentucky. He was by birth a Kentuckian, and in his native State performed a large amount of ministerial work, and with much success. He was ordained in 1832, and for twenty-one years preached to the people among whom he was born. For three or four years he served as missionary, and the rest of the time as an active pastor. Previous to his arrival in Texas he baptized about fifteen hundred persons.

Although past the meridian of life when he came to Texas, he girded himself for his work, and proved to the people of his adopted State that he was one of the strong men in Israel. He served as pastor at Friendship, Collin County, and subsequently at Rowlett's Creek and McKinney. God greatly prospered the labor of his hands. With a clear head and a warm heart he fought sin and preached righteousness among his people, to the great comfort of Christians. His influence was felt in all the surrounding country, and statistics show that many redeemed souls demanded baptism where he labored. He served a number of years as moderator of Elm Fork Association, and was called from his labors to rest with Christ in 1866. From Kentucky and Texas many a child of God will rise, in the resurrection morn, to bless the day that Elder Portman entered the work of the ministry of Jesus Christ.

Elders Myres and Portman have rested from their labors, with many others from the bounds of the association they loved so much, but the Elm Fork Association still lives and annually rallies its forces together around the standard of truth, sending out its messengers during its recesses to declare the way of salvation.

The churches in south-eastern Texas were in much confusion, in consequence of the anti-missionary element and various issues that troubled that section. Many, however, held the true faith and believed in active mission work, and in November, 1849, a convention met with the old Union church, near the town of Nacogdoches, and formed what was then known as the Eastern Texas Association of United Baptists. The old church that had been the rallying point for the old Sabine Association was now the central point for another organization, covering very nearly the same territory. Twelve churches were represented at this meeting: From Smith County, Ebenezer, Tyler, and Harris Creek; from Cherokee County, Salem, Key's Creek, Rocky Springs, and Palestine; from Shelby County, Macedonia, Zion, Concord, and Horeb; and from Nacogdoches County, Union. Elder Robert Turner was the first moderator. The brethren were greatly encouraged by the harmony and success of the new organization, and, at the second session held with Salem church, Cherokee County, there were sixteen churches represented, with five hundred and twenty-five members.

At the fourth session of this body, in 1852, the name was changed to Central Baptist Association. Elder B. E. Lucas moderator, and B. F. Burroughs clerk. Brother Lucas was ordained by Bishop Andrews as a preacher in the Methodist church, in 1843, in the State of Tennessee. He came to Texas in 1846, and settled in Sabine County. In 1850, being greatly dissatisfied with the ordinances and government of the Methodist church, he made application and was received as a proper candidate for baptism by the Hamilton Baptist church, Sabine County, and was immersed by Elder William Britton, in May, 1850. The same year he was ordained by Elders William Britton and Robert Turner. He spent much time as missionary in the Central Association, and served as pastor at

Union, Providence, Mount Moriah, and other points in the east. He afterwards moved to Northern Texas, and his name appears as moderator of Elm Fork Association in 1866.

I have met brother Lucas, and from him learned much of the trials of the brethren in that section, where Moderators and Regulators once spread so much terror and confusion. He has been useful in the past, and still toils on in the gospel of Jesus.

At the sixth annual session, in 1855, the number of churches was ten, with two hundred and ninety-two members. Several had withdrawn to form other associations. Elder John L. Mills was moderator. The body, at this meeting, passed a resolution recommending the "State Legislature to pass the Maine Liquor Law, or one similar," and at the same time urged pastors to speak out decidedly from the pulpit on the temperance question. Whether the churches in that section were cursed with a membership guilty of "dram-drinking" or not, the records do not show. Whether the evil prevailed in that section more than in other parts of the State, I cannot say, but one thing is apparent: those brethren felt that the times demanded them to speak out, and they did it, even in the legislative halls of the country. By this act they unanimously rebuked the devil to his face, and we only regret that there are not more combinations in the land against an evil that creeps like a serpent into the abode of domestic happiness, and, after doing his mischief, mocks at a widow's tears, and laughs at the wretchedness of the fatherless. The spirit of Christianity knows no sympathy with this child, whose father is the Devil and whose mother is Beastly Appetite.

While I do not believe that Baptists in Texas, or in other States, are guilty of this sin above other denominations professing to abide by the principles of Christ, yet the churches do not measure arms with this vice and hurl the monster from their midst as faithfully as the word of God directs. And now let every youth who reads these pages heed the admonition of an old man, and avoid all the paths that lead to a drunkard's life, a drunkard's grave, and a drunkard's hell.

To fill the vacancy in the body caused by the withdrawal of churches at former sessions, new churches were organized and added to the association, and in 1858 there were fourteen churches represented, with four hundred and ten members. The territory extended over the Counties of Shelby, Sabine, Nacogdoches, San Augustine, Rusk, and Panola.

After the decline of my health, in 1847, which caused me to give up my appointment as missionary under the Board of the Southern Baptist Convention, my ministry was somewhat irregular. I kept up the habit, formed in early life, of making long tours whenever I could; but these were not so frequent nor extensive as in former years. Months sometimes passed during which I could do but little travelling, and these intervals in my ministry have continued to grow longer, until I can do but little more than sit in my room and pen the records of labor in the past. When at home, much of my attention was given to the farm, and as so many labor-saving machines had been invented, my mind went in search of an invention that would enable a man to plough up, plant and cultivate more of the rich prairie soil around me than it was possible to do with the implements of husbandry our fathers gave us. I never could see the reason why I should carry a jug of molasses in one end of the sack and a rock in the other simply because my father did, when I could just as easily carry in the same sack another gallon of molasses.

The time to break up new prairie land was upon us, and visiting the shops in the country, I could get no ploughs made, for want of suitable iron. To expedite my work, while the iron was coming, I made a frame in the shape of a common harrow, and put into it five old-fashioned duck-bill colters, which were afterwards increased to nine. The frame rested on two wheels sawed from a post oak, twenty inches in diameter, with a foot to bear up the front end. The whole, thus adjusted, was tied behind the fore-wheels of a wagon and dragged by a long team of oxen. Four acres of land were broken in a day; and when it was ploughed with this same implement across the other way the prairie was thoroughly torn up. This was the first plough on wheels that I had heard anything of, either in Texas or any other State, and was certainly a success.

My friends were greatly amused for the entire season, but this led to the invention of the first planting machine, that I had any knowledge of, the following spring. A plough was attached to the fore-wheels of my wagon, drawn by two horses, that opened a furrow for the corn. The corn was regularly dropped, without the aid of human hands, covered and nicely harrowed over, while the driver rode on the machine and directed the team. Seeing that this experiment was a success, I invited my neighbors to examine the work. As they approached, they inquired of the young man engaged in planting how he was succeeding. His reply was, "I hardly know; but I have certainly multiplied myself into five men. I open the furrow, one; I drop the corn, two; I throw two furrows on it, four; I drag a harrow over it, five; and this is all done with so little labor on my part that I am afriad it is of no account."

Had I gone forward at once and secured patents for what I was justly entitled to, as my friends urged me to do, and devoted a portion of my time to improvements that suggested themselves to my mind soon after that, the result necessarily would have been a large accumulation of money. Some conscientious scruples relative to the loss of time from my ministry and the danger of diverting my mind too much from preaching prevented me until 1857, — at which time I will notice the machine again.

CHAPTER XXVII.

HARMONIOUS ACTIVITY. — 1850 TO 1852.

SO far as I am able to collect information, from minutes and other documents before me, no new association of Baptists was organized in 1850. A sufficient number of general organizations were in existence to meet the demands of the scattered churches, and the combined efforts of the denomination were being put forth to strengthen the infant churches and associations, and to supply the destitute communities with the word of life. While the missionaries, under appointment of the various Boards in the State, were travelling in every direction, and with marked success, there was manifest a general desire to correspond and co-operate, and ministers, with many of the private members of the churches, considered it a privilege to ride a hundred miles on horseback to attend their sister associations, and thus aid, by their presence and counsel, in promoting the interests of our common cause, in every locality. Jealousies and dissensions were strangers among us. By this time we were well agreed in doctrine, and when we differed on plans of operation, exchanges of views, in private and in public, were marked with so much of the spirit of Christ that even these tended to bind us more closely together. Among the happiest recollections of my life is the peace that in those days marked the progress of our Zion. Some of our hearts had been caused to ache, in former times, when errors on points of doctrine were brought in among us, which in their very nature, if received, rolled the apple of discord among the churches that must necessarily multiply itself into a variety of isms and heresies. This state of things was brought about by "false brethren unawares brought in, who came in privily to spy out our liberty which we have in Christ Jesus, that they might bring us into bondage; to whom we gave place by subjection, no, not for an hour." With these we had been compelled to take issue, and "earnestly contend for the faith once delivered unto the saints." Having passed through this ordeal and attained unto unity in doctrine, we rejoiced. Since that time, whenever I have seen petty questions about pet plans and local interests, where no vital principles connected with church polity were involved, sprung upon us, calculated in their vary nature to produce alienation without just cause, the remembrance of the days of our peace, and our fearful responsibility touching the obligation "to keep the unity of the Spirit in the bond of peace," has caused my tears to fall, and my prayers to ascend that God would strangle this child from the "pit," before he grew strong enough to break the bands that bound an active, earnest and loving brotherhood together. As long as we can meet together, and the one sentiment, "one Lord, one faith, and one baptism," finds a united response from the body, there is no excuse for disorganization.

In the midst of the union and concert of action that marked this era in our history, the following resolution was adopted by the Baptist State Convention, at its session in 1850, held with the Huntsville church: *Resolved*, that some suitable person be appointed to collect such historical facts relating to the introduction of the gospel, and the rise and progress of our denomination in this State, as may serve as useful records in coming years; such as the constitutions of all churches and associations, the names and important statistics of all Baptist ministers, and such other facts as he may deem necessary to be preserved. Elder J. W. D. Creath was appointed at the same meeting to perform this service.

According to the spirit and letter of this resolution, brother Creath entered upon the work assigned him in connection with his other arduous labors, and has collected a large amount of material for the historian. By his permission, I have been allowed to have access to this very valuable collection, that has served greatly to refresh my memory concerning some facts forgotten, and I have also been able to add with this assistance some additional facts and incidents to those already in my possession.

During the years 1850 and 1851, except the care of one church near my home, I sustained no relation as missionary to any board, nor as pastor of any church. There was much need of mission work, and, as my health at times prevented me from regular work, I visited the larger portion of the State, west of the Trinity River, at my own expense, and did what I could in strengthening weak churches and preaching among the destitute. While on my way to the Baptist State Convention, held with the church at Independence, in June, 1851, I met for the first time with Elder Jonas Johnston. He had only been a short time in Texas, and being a South Carolinian, had much to say about the old State. As South Carolina was my native State, the right hand of fellowship was at once given in consideration of a common feeling relative to the home of our childhood. The bonds of union that bound us together on that ground were soon forgotten as we entered into conversation on Baptist principles.

Elder Johnston carried with him then, as he has ever done since, the independence of a fearless advocacy of our distinctive principles. God's sovereignty in the plan and in the execution of the grand scheme of redemption, in the exercise of which "he hath saved us and called us with a holy calling, not according to our works, but according to his own purpose and grace, which was given us in Christ Jesus before the world began,'' which doctrine our Baptist fathers boldly declared, finds in brother Johnston a fearless defender and advocate. When he takes hold of this subject, or the final perseverance of all the saints through grace to glory in heaven, or instructs the people concerning the ordinances, government, and perpetuity of the church, he speaks plainly and decidedly. ''If the trumpet give an uncertain sound,'' in the hands of some of our Baptist preachers, on these questions, such a charge certainly cannot be preferred against Elder Johnston.

He has served as pastor of several churches in Grimes, Montgomery, and Walker Counties, has baptized a great many people, and has always taken an active interest in the mission and educational enterprises of the denomination in Texas. But few of our Baptist preachers have possessed financial resources equal to his, and whenever and wherever the cause has demanded money he has given clear evidences of large benevolence.

In the very heart of the territory in Eastern Texas, from which the Cherokee Indians were driven out, in 1839, by the Texan troops, under Douglass, Burleson and Rusk, the voices of Elders W. H. Ray, J. Rasbury, and others, for some time previous to 1851, had been declaring the way of salvation to the people in Smith and Rusk Counties. In December, 1851, messengers from three churches — Harris Creek and Mount Zion, Smith County, and Sharon church, Rusk County — met with the Mount Zion church and organized the Cherokee Baptist Association. Elder Rasbury was elected moderator, and Elder Ray clerk, and they were the only ministers present at the organization.

When either of these churches was organized I cannot state definitely. At the first session, Elder William H. Ray was appointed as missionary, and at the second session reported five months' work performed. His salary was fixed at four hundred dollars per annum, and his report shows that fifty dollars of the amount due were paid by the Baptist State Convention. At this session, held with Harris Creek church in September, 1852, resolutions were adopted favoring ''the establishment of a Female High School at Tyler, Smith County.'' Earnest efforts were put forth by these brethren in the educational and mission causes, but, owing to the scarcity of minutes at command, I can give no satisfactory statements as to their progress, except that at the fifth session, held with Carmel church, Smith County, in 1855, with Elder J. S. Bledsoe moderator, the number of churches was fourteen, with seven hundred and three members. The territory at that time embraced the counties of Rusk, Smith, Wood and Van Zandt.

In the month of December, 1851, Elder Thomas Chilton came from Alabama to Texas, and settled as pastor of the church in Houston. I first met him in the Union Association, held with the Montgomery church, in October, 1852. He was a man of acknowledged ability, much decision of character, and as a pulpit orator ranked among the first in the denomination. His personal appearance was commanding, and his manner bold and fearless. With a clear head and an earnest delivery, he pressed his conclusions with great power.

His name appears as pastor at Houston for two years. On the sixteenth day of August, 1854, while serving as pastor of the church at Montogmery, he died in the modst of his flock, who loved him much. His sojourn among us in Texas was short, but his name and deeds of love live on, and will, in the memory of many Texas Baptists.

Perhaps never in the history of the country was the husbandman better rewarded for his labor than in 1852. There have been years before and since when God seemed to speak to his people in Texas as he did to the ancient Israelites: "Thy heaven that is over thy head shall be brass, and the earth that is under thee shall be iron." During that year the rains from heaven watered the earth bountifully, and the land brought forth an abundant harvest, that greatly encouraged the people, and swelled the tide of immigration that steadily poured into the State from every quarter.

The brethren were able to provide more liberal means for the cause of missions and for the promotion of our educational enterprises. God blessed the churches in many localities with refreshings from his presence, and our minutes show ingatherings both by baptism and by letter.

The Bethlehem Baptist Association was organized at Woodville, Tyler County, on the twenty-fifth day of September, 1852. Messengers from five churches, Sardis, Indian Creek, Zion, Providence and Bethel, were present. Three ordained and two licentiate ministers were present, and the total membership represented was eighty-eight. Elder E. Vining preached the introductory sermon and was elected moderator.

He was by birth a Georgian, and was ordained as a minister in 1846, at the age of thirty years. In 1847 he removed to Florida, and in 1850 came to Texas, and settled in Jasper County. With Elder Vining I had no personal acquaintance, but from a circular letter, read before the first session of the association, clear evidences are given that he was a man of ability, and hesitated not to avow his principles boldly. His position among the churches and in the association shows that he had the confidence and appreciation of his brethren among whom he labored. In 1855 he was pastor of four churches in this body, and was moderator of the association at every session from the organization till his death. He died March 5, 1856, at his residence in Polk County. The light which he borrowed from the Redeemer shined in Texas for six years, and then the Master called him home.

At the second session, in 1853, held with Providence church, Jefferson County, six queries were presented for discussion, and according to the word of God were correctly answered. While these questions were of a general character, and answers correctly given, I must be permitted here to record my conviction that a vast amount more of trouble than good has grown out of the discussion of and answers to queries brought before our associations. Especially is this true when queries of a general character are presented, involving cases of a local character. In my early ministry questions of this kind were frequently forced upon us in our general meetings.

One incident will serve to illustrate this evil. In 1832, in the Obion Association, in the State of Tennessee, where there was much opposition among Baptists to masonry, a query was presented on this subject that came very near rending the body asunder. The moderator of the association, I, and others, were members of that ancient order, and at peace in our churches on this question. Our cases brought the query up. This association, after an animated discussion, threw the query out, and wisely refused to give any answer at all. The better plan, when exceedingly difficult questions involve the harmonious action of a church, is, to call a council from sister churches, and allow the case to be decided by disinterested brethren in the locality where the difficulty occurs. Queries sometimes may and ought to be sent to associations; but the parties who send them and the bodies who entertain them should be exceedingly careful, or the result in many instances will be evil instead of good. The main object in an association should be to effect by combinations in the work of evangelization what cannot be effected by single churches; and while such organizations should labor to promote unity in the faith of the gospel, questions that "gender strifes," rather than the "peaceable fruits of righteousness," should always be avoided. The reader will bear in mind that there is no intention upon the part of the writer to cast stones either at the brethren who introduced or those who entertained the queries alluded to. They were all of a general character, and the answers were correctly and properly given.

At this same session an executive committee was appointed to labor in the cause of domestic missions during the recess of the body. At the third session, in 1854, held with the Sardis church, Newton County, the missionary, Elder E. S. Phelps, made the following report: "I have been employed by the Executive Board, as domestic missionary, one hundred and thirty-nine days; travelled two thousand four hundred and eighty-two miles; preached ninety-two sermons; delivered nine exhortations; visited eighty-four families; baptized seven, and assisted in the constitution of two churches."

The name of Elder Reuben E. Brown appears as the missionary of this body in 1856. He was employed as missionary in May, and on the first of November reported, as the result of six months' work, one hundred and eighty-eight baptisms, and one hundred and eighty-six sermons preached. During this time he aided in the constitution of three churches, the ordination of one minister and seven deacons. He had been in Texas only a short time previous to this appointment.

Elder Brown was by nature an extraordinary man, and all who have ever associated with him are bound to admit it. I met him frequently during his sojourn in Texas. He was from Alabama, and his first report, just recorded, shows that he was a revivalist. He labored for a number of years as a Methodist preacher previous to his union with the Baptists. Although a man of limited education, he was wonderfully gifted with ability to move the masses to an earnest consideration of things eternal. In person he was very tall, and, like Saul, "from his shoulders and upwards he was higher than any of the people." His voice was clear as a trumpet, and of great strength and endurance. In sacred song he had but few if any equals, and frequently melted large congregations to tears under the strains of music from his single voice. His was the gift of exhortation promised to the churches, and, after a life of usefulness, he died at his post as a preacher in the city of Galveston, during the late war, in hope of a blessed immortality.

The brethren of this association prosecuted their work with earnestness and zeal. Harmony and activity marked their operations, and at the sixth session, held with the Beach Creek church, Tyler County, in 1857, messengers came up from twenty-two churches, with a total membership of six hundred and forty-two. Their territory at this time embraced seven counties, Jasper, Newton, Orange, Tyler, Polk, San Augustine and Jefferson. At this session the name of Elder W. B. Prewett appears as pastor of the church at Moscow, Polk County.

He was by birth an Alabamian, and moved to Texas in 1850, at the age of twenty-three, and settled in Trinity County. He was baptized by Elder J. V. Wright into the fellowship of the Bethel church, Polk County, and subsequently ordained to the work of the ministry. With this young brother I met but once, in one of the sessions of the Baptist State Convention, but the impressions made upon my mind by this interview have not been erased. His piety was deep and ardent, and his mind was intent upon the investigation of the word of God. He gave the clearest evidences of deep humility, combined with fixedness of purpose and consecreation to his work as a preacher.

On the twenty-second day of March, 1859, at the age of thirty-two, God called him to rest from his labors.

While we rejoice at the prosperity that attended the labors of both ministers and churches in the bounds of Bethlehem Association, it is a painful duty thus to record the death of three of her most useful preachers, whose lives were connected with her early history.

CHAPTER XXVIII.

THE EASTERN CONVENTION. — 1853.

FANNIN and Grayson Counties join the Indian Territory, and are near the centre of our northern border, and a little east of the head waters of the Trinity River. The banner of the cross was borne through that region by Elders Harris, Walker, and others; men and women were converted and baptized, and these, with others who had been baptized in the older States, rallied beneath the flag on which was written "One Lord, one faith, and one baptism," and were organized in Baptist churches.

On the twenty-fifth of June, 1853, representatives from three churches, Pleasant Hill, Salem, and New Hope, met in convention at Bonham, Fannin County, with Elder John O. Walker moderator. These messengers then and there declared their belief that it was expedient to form an association. In October following, messengers from four churches, with a membership of one hundred and forty-five, met with the Pleasant Hill church, Grayson County, and organized the Sister Grove United Missionary Baptist Association. A Missionary Board, consisting of only three brethren, S. D. Rainey, Gideon Smith, and Z. Ray, was appointed at this meeting, and at the second session, held, in September, 1854, with the church at Bonham, Elder B. Watson moderator, said committee made the following report: —

"The Executive Board of Sister Grove Association beg leave to report, that they employed our beloved brother, J. Briscoe, at the rate of fifty dollars per month. Commencing his labors on the first of June last, he reports that, with the assistance of other brethren in the ministry, the good Lord has abundantly blessed the feeble efforts of the humble instrument thus employed. He has travelled six hundred and thirty miles, preached fifty sermons, delivered thirty exhortations, witnessed one hundred and twenty conversions, and baptized eighty-five persons." In October, 1855, the association met with Ephesus church, Choctaw Nation, Elder T. J. Harris moderator. Eight churches petitioned for membership and were received at this session.

The report of the board shows that Elders Briscoe and McCombs were the missionaries. Three active men on an executive committee are far more effective in carrying forward either mission or educational enterprises than a committee of a larger number ordinarily. Here is a clear case: Three brethren received a commission form the association, in 1853, to look after the mission work, and they did it well. Encouraged by their success, and with a view of enlisting the churches more earnestly in the good cause, the body passed a resolution, in 1855, appointing an executive board of fifteen, one from every church, instead of three. As is usually the case, it was difficult to get a quorum, and the many trusted to a few, who did not feel disposed to assume responsibilities for the many. Under this system the work was greatly embarrassed. At the next session, in 1856, held with Concord church, Hunt County, the number of the board was reduced to three, and the mission cause moved forward again. In view of the many failures that have befallen us in consequence of these large committees, surely it is time the number was reduced to three, or five at most, and if we can't find a few brethren that can conveniently meet together, to whom we can confide any interest we have, the sooner we give up the enterprise the better. The appointment of a few means work, while the appointment of many means honorary membership, which class of members the Baptists of Texas can well do without.

This association, in 1854, recorded the sad intelligence that Elder John O. Walker was dead. The following is an extract from the minutes: "The Lord in his providence has taken from our midst, by death, one of our most faithful and efficient ministers of the gospel, brother John O. Walker, who was truly an humble and pious follower of Jesus Christ, and who had, according to his acquaintance, as many friends and as few enemies as perhaps ever falls to the lot of man on earth. His religious life was characterized by charity to the poor, sympathy to the afflicted, kindness and hospitality to all classes of society. He was the friend and advocate of all benevolent institutions, and was ever ready to labor in the advancement of his Master's cause, in any and every way that it was possible for mortal man to toil." The first moderator of this band of noble workers passed to his reward, but his influence was felt by others, whom God raised up to carry the work forward.

In 1858, when the number of churches was twenty-five, with about twelve hundred members, a committee of five was appointed "to select a suitable location for the establishment of a Denominational School." In 1860, the committee reported the Ladonia Male and Female Institute, located at Ladonia, in Fannin County. Gideon Smith, the moderator that year, was appointed president of the board of trustees. The following year the school was reported in a prosperous condition, under the charge of Elder J. C. Averitt and lady. It subsequently passed into the hands of Elder W. B. Featherston as president, who holds the position still.

As an educator Elder Featherston ranks among the first in the country. By his indomitable perseverance, and the co-operation of his brethren, this school has been made a great blessing to the people of northern Texas. Baylor University has the honor of lending aid to Elder D. B. Morrill as one of its first beneficiaries. His voice was heard in the defence of the truth, north, east, south and west, and some of the last labors of his life were given to the people at Ladonia. Elder Featherston was his bosom friend and earnest co-worker in defence of the truth, and when this good man and self-sacrificing preacher passed away, leaving a widow and a large family of children, without earthly riches, he engaged to see that these children were educated. While the friends of Baylor University rejoice in being permitted to have part in the education of the father, Ladonia and the noble-hearted president, with the brethren who sympathize with him and aid in the work, have still greater reason to rejoice while engaged in that which is even more acceptable to God. He who lends the aid of but a single farthing in the education of the children of such a man, with the fear of God before his eyes, cannot fail to reap his reward.

Among the names of ministers in this body is that of Elder S. J. Wright. When but a boy, and as early as 1839, I knew him on the Colorado River. His father and two older brothers were preachers, and when God gave him a new heart, his mind was fired with a desire to publicly point out the way that leads to God. Not satisfied with his mental attainments, he entered Baylor University after he was the head of a family, and there cultivated his mind with the same earnest efforts that subsequently marked his ministry. He afterwards moved to northern Texas, and served as pastor in Fannin and Garyson Counties until his death, which occurred on the fourteenth of October, 1868. He held a high position in the estimation of his brethren, and did much to lay broad and deep the foundation of that success that has attended the cause of Christ in that section.

At the ninth annual session, in 1861, the three churches that were represented at the organization were increased to thirty-two, with a membership of fourteen hundred and forty-three. The territory at that time embraced the counties of Fannin, Grayson, Collin, Denton and Hunt.

As has been seen, the Baptists east of the Trinity River increased rapidly both by immigration and by baptism, and according to the judgment of the leading spirits in that section the territory was too large for one general organization. In accordance with this view, a convention met at Larissa, Cherokee County, in November, 1853, and formed The Texas Baptist General Association. The introductory sermon was preached by Elder M. Lepard. He had then but recently entered the State. As a preacher he was earnest, and a bold defender of the principles of that sect everywhere spoken against. The churches in Rusk County felt the power of his ministry only a short time, until a cancer claimed him as its victim. While absent from his family, in Tennessee, seeking medical aid, he passed to the upper sanctuary, in January, 1859, and rests from toil and suffering.

The first president was Elder I. H. Lane, from Cherokee County, who after a long and successful ministry fell asleep in the arms of the Master he loved so much, in April, 1858. The constitution under which this body was formed was very much the same as that adopted by the Baptist State Convention. The second article reads as follows: "The objects of this body shall be missionary and educational, the promotion of harmony in the denomination, and the organization of some general system for the advancement of the Redeemer's kingdom."

On the twenty-fourth of May, 1855, this organization was dissolved at Tyler, Smith County, and the Baptist Convention of Eastern Texas organized upon the spot, with Elder Wm. H. Stokes, president, and Wm. Davenport secretary. At this meeting the difficulty between Elder G. G. Baggerly and brethren in the west was settled. Differences of opinion had existed as to the proper appropriation of denominational funds, and charges of unfaithfulness were rashly made by Elder Baggerly against the Board of the convention in the west. After a rigid examination on the part of a judicious committee, receiving testimony from brethren east and west, the apparent discrepancies were all accounted for; and when their report was made to the convention, a resolution was passed declaring the differences "amicably adjusted." As an evidence of their determination to heartily co-operate with the convention in the west, the "Texas Baptist" was adopted as their denominational organ, and Elder Wm. H. Stokes was elected corresponding editor.

Right earnestly did these brethren now address themselves to their work. Their missionaries penetrated many destitute fields, and rendered efficient aid in establishing the struggling churches. The voices of Elders Tucker, Witt, Clemons and Morrill — all able, earnest men of God — were heard throughout the territory of the east, as general agents of this body, pleading the cause of missions among the churches.

In 1859, when the convention was held with the church at Bonham, Fannin County, with Elder A. E. Clemons president, a committee of fifteen was appointed to "take into consideration the propriety of building up a denominational school of such a character as will meet the wants of the denomination in eastern Texas." In 1860, when the convention was in session at Tyler, Smith County, with Elder J. S. Bledsoe president, the East Texas Baptist Male College was located at Tyler, and placed by the board of trustees in charge of Elders W. B. Featherston and J. R. Clark. Before this institution of learning was thoroughly organized, and before its friends had time to rally in their strength around it, the war between the States was upon us, and ere the struggle was ended the school ceased to exist.

Prominent among the names of ministers in this body appears the name of Elder John H. Rowland. With a piercing blue eye, angular features, and well-developed head, he gives evidence of a man of mark. Although he did not enjoy the advantages of early education, the superiority of his natural powers of intellect, brought to bear upon the word of God from the pulpit, causes his audience to forget the defects in his early mental culture. Possessed of a full and commanding voice, with clear and vigorous thought, he presses truth upon the minds of the people with great power.

He came from Mississippi to Texas in the spring of 1853, and, passing in a boat up the Trinity River, landed on the soil of Anderson County. In a strange land and out of money, he had a fine opportunity to manifest that independence of thought and action that has been characteristic of the man in all his subsequent history in Texas. He entered a store in the town of Palestine, and spent a few weeks in keeping books. As he came to Texas to preach, and not to keep books, the position was by no means a pleasant one; and receiving a pressing invitation from brother John Smith to pay him a visit five miles in the country, with a prospect of finding employment as a preacher, he left the store and the town to preach among the destitute. He left the town as he entered it, afoot, and walked to the house of brother Smith. Here he was furnished with a horse, and from this brother received a pledge, that, if he would devote his time and energies on that destitute field, he should not only be provided with a horse, but should in addition to this have food and raiment.

The surrounding country was then one vast field of destitution. No sooner did this missionary enter that field than evidences were given of the divine approval of the man and his work. One revival after another, at different points in the country, followed in quick succession. Before the year closed more churches were organized than he could supply. Among the first work he did was the baptism of two of brother Smith's children, and as he returned with the son of sixteen summers from the water, he said to the father, "I feel that I have baptized a preacher." That son was Elder M. V. Smith, whose name is familiar to the Baptists of Texas. With the exception of a short interval, during which time he served as pastor in Freestone and Limestone Counties, his ministry has been confined to the east, among the churches of Anderson, Smith, and Rusk Counties.

It was my privilege to meet Elder Rowland for the first time at the Trinity River Association, in 1856, held with the Springfield church, of which he was at that time the pastor. With him I afterwards spent a week among his people. During these interviews, impressions were made upon my mind and attachments formed for the man that are among the most pleasant reminiscences of my life. He is of that class of men, outspoken and impulsive, who when right are "mighty right," and when wrong are "mighty wrong."

In him the Baptist cause has ever found a bold and earnest defender. Almost twenty years with him have passed in Texas, during which his influence has been felt over an extensive field. He is now the pastor of the church at Tyler, Smith County.

The Baptist Convention of Eastern Texas, at its session in 1863, inaugurated the "army mission" in the trans-Mississippi department. The churches, instead of banding themselves together, and selecting and sending to the field their most prudent and pious preachers, — men adapted to the work, — trusted to government officials to supply the army with chaplains. These officials were mostly men of the world, and without the fear of God before their eyes. The selection of chaplains in a large number of instances was a matter of mere personal favoritism. Two years of war passed, and most of the preachers who filled this position in the early part of the war had returned to their homes discouraged. The brethren in the east now had their attention directed to the fact that the commission to preach the gospel was not committed either to human governments or human organizations, but to the church of God. A man was found whose judgment and heart were in perfect sympathy with this scriptural policy, and who was willing to consecrate himself to mission work on the tented field. Believing that the church and State were organized for separate and widely different purposes, his conscience stumbled at the union of church and State, even touching the chaplaincy. He was willing to serve the churches as a missionary, but was unwilling to be trammelled by the secular power in the exercise of his ministry. That man was Elder M. V. Smith. It was my privilege to hear him on this subject before the Baptist State Convention, at Huntsville, during the war, by request; and it was one of the happiest efforts of his life. The Scriptures and personal experience and observation were brought so forcibly to bear upon this great question, that the Baptist heart could but respond a hearty amen to the utterances of the speaker.

At the session of the Eastern Convention alluded to he was appointed as missionary to the Texas armies, then in Arkansas and Louisiana. Being present, the appointment was accepted, provided his resignation as an officer in the Confederate army was accepted.

Elder Smith was by birth a South Carolinian, and was brought by his parents from Mississippi to Texas in 1850, at the age of thirteen years. Anderson County was the home of this family for about nine years. Although the subject of deep religious impressions from early childhood, he did not realize a change of heart until sixteen years old, under the ministry of Elder J. H. Rowland, by whom he was baptized in July, 1852. The impression fixed upon the mind of the administrator of the holy ordinance, that the boy would become a preacher, was in perfect keeping with impressions soon made upon the heart of the youth by the Spirit of God. The handwriting of God was so plainly written and sealed by his Spirit, that the people soon recognized the fact that he was "a chosen vessel" to bear the name of God before his neighbors. That spirit of love divine that brought the Redeemer to the earth for the accomplishment of his glorious work on behalf of a world already condemned, in consequence of unbelief, had so thoroughly quickened the dead faculties of the soul as to bring forth fruit unto holiness, and as he had "freely received," he was at once impressed with his duty to appropriate his talents — all — to the service of God.

His father had wisely taken the delicate child from his studies at the age of eleven years, and steadily kept him at labor on the farm for five consecutive years. Having determined to consecrate his life to the Christian ministry, he entered school in September, after his baptism in July. His mind thirsted after knowledge, and with him there was no time to lose. Great was the work, and the responsibilities high as heaven, deep as hell, and boundless as eternity. Although he was physically weak, God blessed him with a vigorous mind, capable of grasping knowledge rapidly, and possessed of great powers of concentration. He commenced exercising his gift, in a modest way, immediately. In 1855, at the age of eighteen, he was licensed to preach by the church at Palestine, while still pursuing his studies. In consequence of the great scarcity of ministers in that section, the brethren soon suggested the importance of his ordination. His extreme diffidence and the consciousness of responsibility caused him to beg for time. Hiding himself behind the injunction of Paul, "Lay hands suddenly on no man," and often repeating to himself, "Who is sufficient for these things?" he induced the brethren to defer his ordination, until a request came from a church in an adjoining county, when he submitted to the hands of the presbytery, in 1858, composed of Elders G. W. Baines, D. B. Morrill, J. R. Malone and N. Crain.

He was immediately called to the care of the church at Palestine, and others in that vicinity, and with all the means at command continued in school until 1859, when he removed to Rusk County, and took charge of churches in Smith and Rusk Counties. He now lived in the family of his former pastor, J. H. Rowland, at whose hands he had received the ordinance of baptism, and the two felt much towards each other as did Paul and Timothy, the father and son in the gospel. An amicable arrangement was made, by which Elder Rowland was taught in the languages, and the young preacher received the benefit, in return, of Elder Rowland's knowledge of theology and pastoral work.

In sympathy with the tide of southern patriotism that swept over the country, Elder Smith entered the army, and accepted a captain's commission. It was an evil day with him when he left the pastorate and assumed the duties of a soldier under Caesar; but, like many other preachers, he drifted with the mighty current, and for nearly two years served as a soldier, with honor both to himself and the country. A merciful God overruled it for good. I frequently heard from this young brother through the passing soldiers, and, although I had never seen him, had learned to love the captain, who, after long marches at the head of his infantry company, preached at night, and on all occasions when his duties as an officer would permit.

At length, while the writer was pastor of the church at Bedais, Grimes County, in 1863, I was introduced to Captain Smith, at the time and place appointed for Saturday conference. I insisted that he must preach; but he remonstrated earnestly, saying that he had just ridden five hundred miles on horseback, — had been detained on the way by sickness, — had stopped twice to preach, — and had heard nobody preach but himself in over a year. Those who knew him will remember how earnestly he can beg; but in this case no excuse was taken.

I may remark that I sympathized with the young man in the midst of his embarrassment. I was aware of the fact that the silken cord of love bound his heart and the heart of a pious young lady, a member of my flock, very tenderly together, and that upon this mutual affection the *pledge* had been made. All unknown to him, while in the camp in Arkansas, Miss Cornelia Camp, the daughter of John and Eliza Camp, had permitted me in confidence to read some extracts from his letters. The family were present, and neither they nor any member of the congregation had ever heard him preach. Notwithstanding all this, we pressed him into the pulpit.

The church at the time was in trouble, of which he was entirely ignorant. Reading the eighteenth chapter of Matthew, and some other passages of Scripture bearing upon the same subject, the theme selected was *church discipline.* In his treatment of this subject he showed an acquaintance with the Scriptures that would have honored riper years. Some of the more intelligent members charged the writer with posting the preacher, owing to the direct, bold and independent bearing on the cases under the discipline of the church at the time. On this occasion the secret of his success as a preacher was plainly manifest. He was earnest, plain and practical. His sermon on Sunday, touching the cross of Christ, was characteristic of those burning, pathetic appeals that have melted into tenderness so many hearts in the army and out of it.

If a few days it was my privilege to pronounce the marriage ceremony between him and Miss Camp. After a short stay with the loved ones, he left for his post in the army, and on his return attended the Eastern Convention, of which he was a member, and which met at Tyler. By this body he received the appointment before alluded to. His resignation was cheerfully accepted, and all his energies were devoted to the mission. No sooner did the churches undertake this work than the divine approbation, without a doubt, rested upon the enterprise. Soldiers by hundreds were converted and baptized. Elder — McCraw, who had long served as a private soldier in General Walker's division, was assigned to duty as a chaplain, and by his earnest ministry influenced a number of his associates to accept the purchase made by the Captain of our salvation. He was at this time only a licentiate preacher; but, in consequence of the success that attended his preaching, the church at Ebenezer, Walker County, of which he was a member, gave authority for his ordination in the army, and requested the ordained ministers on the field to form a presbytery and set him apart by the imposition of hands, in the name of Christ, to the holy office of a Christian bishop, according to the teachings of the New Testament. The presbytery consisted of Elders M. V. Smith, A. L. Hay, and W. A. Mason. Up to this time he was compelled to call upon others to administer the ordinance of baptism to the numbers that repented under his ministry. After this he baptized a large number himself, and continued in the faithful performance of his duties till the close of the war.

The missionary of the Eastern Convention soon baptized and organized into a regular Baptist church over a hundred soldiers, and still the work went forward. Elder J. F. Johnson was soon afterwards sent from the east, and Elders F. M. Law, J. V. Wright, and W. A. Mason, from the west. Eternity alone can reveal the amount of good accomplished by this noble band of Texas preachers.

The delicate constitution of the first missionary was at times prostrated by the exposure of preaching and immediately afterwards falling asleep with heated lungs upon the open field. As his strength returned he was at his post again, till the very last; and a large number of men bore certificates of baptism at his hands to various churches throughout the State, and some of then are now preaching the glorious gospel of the blessed God.

Old soldiers say that the army is the place to try men's souls, — and so it is. Some men who stood high in the estimation of their brethren, and who bore the office of the Christian ministry, sank beneath the waves of trial; but this man came out of the army without a stain upon his ministerial character, having passed, under the providence of a merciful God, through it all. As much as he loved his brethren in the east, and as much as he desired to return to his old field after this mission was ended, Providence ordered otherwise, and in 1865 he settled at Navasota as pastor of that and two other churches in Grimes County. During the two years that he occupied that field he was compelled, in order to support his family, to gain a part of his living from other pursuits. He labored one year in the schoolroom, but was driven from this by his declining health. The same necessity still existed, in consequence of the meagre support given him in the ministry; and for a year, still serving the same congregation, he labored as a carpenter in the town of Navasota.

In 1867 he was called to Washington County, and served the churches at Brenham and Chappell Hill one year. In 1868 the church at Brenham asked for his whole time, and, in keeping with his desire to be the pastor of one church, he accepted. God has wonderfully blessed him in this pastorate. Many have been added to the church under his ministry, and it is one of the most efficient bodies in the whole country, ready to every good work.

This church was organized by Elders R. E. B. Baylor and H. Garrett, in December, 1846, with nine members, about four miles north of Brenham, and was then called New Year's Creek church. Elder David Fisher was its first pastor.

More than five years have passed since the present pastor assumed this charge, and every month of this time has seemed to bind pastor and people more closely together.

In addition to the arduous work of a pastor in the midst of a large congregation, he is the corresponding secretary of the Union Association, and corresponds extensively with about thirty-five churches, and their agents appointed to collect for the missions. As I have seen large packages of letters being sent to the churches repeatedly, I have been forcibly reminded of the empty reports of a number of corresponding secretaries, and am forced to the conclusion that the brother who accepts this office ought either to stir the churches up to the performance of their duty in the mission cause, or resign.

Elder Smith is now in his thirty-fifth year, having passed through sixteen years of an eventful ministry. In person he is about five feet ten inches high, erect and well formed. His head is well balanced, with reverence, veneration and firmness full, — hair black, — a laughing, penetrating eye. In him the social element largely predominates. His voice in the pulpit is clear and distinct, but not strong. His head has ever been full of waters and his eyes a fountain of tears when he touches the plan of salvation and the sinner's hopeless state out of Christ.

Much of his usefulness for the past nine years is due to his wife, who, according to the word of God, fills her position well. She is vigorous, watchful and pious. But little does she allow domestic cares to trouble him, whose thoughts she would have wholly given to the work of the ministry. The time she closely watches, and often reminds him of the importance of meeting engagements promptly. Although full of spirit and capable of the deepest feeling, she bridles well her tongue, — and God have mercy upon the poor Baptist preacher whose wife yeilds to tongue and temper in the midst of his flock.

At Larissa, Cherokee County, the Judson Association was formed in November, 1853, with seventeen churches, containing about eight hundred members. Elder I. H. Lane was the first moderator, and James E. Teague the first clerk. This is the largest number of churches that we have yet noticed represented in the organization of any association. As evidence of their aggression, at the fifth session, in 1857, held at Crockett, Houston County, there were thirty-six churches, with about fourteen hundred members, scattered over the territory embraced by Anderson, Houston, Cherokee, Rusk, Henderson, Nacogdoches and Trinity Counties.

CHAPTER XXIX.

REVIVALS. — 1854 and 1855

THE year 1854 witnessed a large amount of labor upon the part both of the ministry and churches. No new associations were formed; but; while all were satisfied that the existing organizations met the demand, the work was prosecuted vigorously in every quarter. Far up the Trinity the older soldier, Elder N. T. Byars, with Elders J. C. Hunton, E. A. Daniel and others, was organizing churches and preparing the way for the West Fork Association, West of the Brazos, Elders J. S. Allen, W. B. Eaves, J. G. Thomas and others, were forming churches, soon to be banded together in the Little River Association. Two general organizations, that were very soon acting in perfect harmony, were sending their agents over the entire territory, developing rapidly that unity of faith and practice that is till this day the joy of the few of "the old guard" that still live. Two such organizations have up to this time met the demands of the denomination, notwithstanding the Texan territory is so large. The Eastern Convention ceased to exist, since the war, but the General Association has been organized in its stead, and embraces a number of churches in Northern Texas, in addition to those in the east. As near as I can ascertain at this time, with the minutes of all the associations that were then in existence before me, the number of Baptists in Texas, in 1854, was about ten thousand. Surely God had wrought wonders on this field, in eighteen years, and our joy was inexpressible.

During this year, the "Hero of San Jacinto" appeared upon the field again; not to drive the Mexicans and Indians from the soil of his adopted State, but to enroll his name among the believers in Christ and lend his influence in extending the conquests of religion. In November, 1854, he presented himself as a candidate for baptism to the church at Independence, and after a few simple statements as to the change God had wrought in his heart, he was approved by the church as a proper subject for baptism. On the nineteenth day of the same month he was buried in baptism by Elder R.C. Burleson, the pastor of the church. It was his delight afterwards to attend our general meeting, whenever his official duties would permit, and give the benefit of his counsel to his brethren in the mission and educational enterprises of the denomination. His speech on one occasion before the Baptist State Convention on the Indian mission was one of the masterly efforts of his life, and did ample justice to his reputation as an orator.

He remained a consistent member of the church until his death, in the town of Huntsville, on the twenty-sixth of July, 1963. It was my privilege to visit him a few days previous to his death. Calmly and deliberately he spoke of the passage he was about to take across the river, and expressed the strongest confidence in Christ. Thus General Sam Houston passed away, whose memory so many of us love to cherish.

So soon as the clouds of winter passed away, and the warm sun of 1855 caused vegetation on Texas hill and prairie to manifest its life again, the "Sun of Righteousness" began to warm in an unusual manner the hearts of his people in this latitude. Evidences of a revival spirit were manifest in the early part of the season, and, as the summer came on and September drew near, showers of grace from the clouds of mercy that hung around the mercy-seat of the upper sanctuary fell copiously upon our Master's vineyard, and a glorious harvest was the result. From all quarters glad tidings came that souls were born to God. The Lord specially showed his power and willingness to save wherever the gospel was preached on fields formerly destitute of the word of life.

When the Trinity River Association met in September with the Little River church in Milam County, thirteen new churches petitioned for membership and were received. The revival spirit pervaded the whole body during the entire session, and for a week after the association adjourned the people of that community continued to wait on the Lord with abundant manifestations of his presence to save sinners.

This was one of those dry seasons that in years past visited this country. Water was so scarce that it was necessary either to disappoint the brethren expecting to attend the meeting, or pitch our tents in another locality. Accordingly, the brethren camped at the Block House Springs, seven miles from the church edifice, in primitive style. This was the beginning of a Baptist camp-meeting. Preaching began on Thursday night, and the first service was full of interest. Friday the association convened, and all through the day the plainest indications were given that the Lord was among the people. The committee on preaching did not stop to go through the whole list of preachers present, and, lest they might hurt somebody's feelings, make arrangements to give every man a chance; nor did they take into consideration the fact that if some of the more prominent brethren were put forward at the beginning, or not invited to preach at times when the congregation would be larger than others, they might go away dissatisfied. The only question seemed to be, to find the man whose heart was full of the Spirit of God, and the most likely to effect, under God, the greatest amount of good.

Elder R.C. Burleson was chosen as the man for Friday night. Selecting his text from the book of Numbers, he read with emphasis, "And be sure your sin will find you out." My opinion is that he has seldom in life excelled that sermon. Sin was held up to the gazing audience, deceiving first its votaries, then causing its subjects to openly violate, step by step, both the laws of man and God. The judgements of God, that will certainly be measured out to evil-doers, thereby showing that "the way of the transgressor is hard," were so forcibly presented, that sinners cried for mercy and fled to Christ for deliverance.

The association adjourned on Tuesday, and on the following day more than thirty persons had been buried in baptism. Every one who had given public demonstration of a decided interest professed conversion and was baptized. Under these circumstances, the writer delivered a farewell address, supposing that it was the mind of the Spirit to close the meeting. Earnest solicitations were now sent to continue the services, and at night a large number of persons came in from a distance. After the sermon, about thirty-five of these new-comers presented themselves as penitents, inquiring the way of life. Services were continued until the following Monday, and in all sixty-four persons were baptized.

On the twelfth of October, representatives from twelve churches, with three hundred members, met in convention with the church at Birdville, Tarrant County, and organized the West Fork Association, with Elder N. T. Byars moderator. Eight ordained ministers took part in the work. Two years later, when this body met with the Little Bethel church, Dallas County, in 1857, there were twenty-one churches, with six hundred and thirty-three members.

Along Little River and its tributaries, in the Counties of Bell, Williamson, Milam, and Burleson, there was a body of earnest, working Baptists. Among these Elder W. B. Eaves was justly entitled to the pre-eminence as a revivalist. He had previously been ordained by request of the old Providence church, Burleson County. He was from Alabama, and, like many others, a real Johna, tried to avoid a work that the Lord intended he should do. For a number of years, in the old State, his mind was exercised with reference to the ministry. He was ordained deacon, but even refused to offer public prayer lest he might be induced to exercise further his gift. Finally he moved to Texas, hoping in a new country to relieve his mind of the burden.

Shortly after his arrival, he spent a night at my house in search of land, his church letter still in his trunk. Before retiring I asked him to read a chapter and pray, having learned from his associates that he was a Baptist deacon. Having never done such a thing out of his own family, he begged to be excused. But it never was my habit to allow such men to kick out of the traces, and I pressed him into service. As he read and sang and prayed, his voice trembled; but behind the embarrassment I clearly saw that his soul was burdened under a sense of duty left undone. Before parting next morning, I sought a private interview, and declared to him the impressions upon my mind. His convictions were now intensified, and coming under the leadership of old Deacon Pruitt, he was soon in the midst of a revival that developed his gifts.

Along both sides of the Brazos River, north of Washington and Grimes Counties, as far up as Waco, the voice of this man was heard for many years, calling sinners to repentance. He served both as pastor and missionary, and but few men were more successful. For the past five years, in consequence of his shattered health and poverty, he has devoted much of his time to secular pursuits. During the past year, while out boring wells during the day, not far from the city of Bryan, he preached at night, and baptized about sixty in one community.

The Little River Association was organized in the town of Cameron, Milam County, on the ninth of November, 1855. Eleven churches were represented, with a membership of five hundred and sixty-five. Four ordained and six licentiate ministers were members of this body. Deep and earnest piety pervaded this brotherhood, and in answer to prayer the Lord not only saved sinners, but manifested his Spirit in firing the hearts of the rising membership with great devotion and activity. God has given to these churches a large number of young preachers. The minutes of no association show so many licentiate preachers in the beginning. Four out of six of the licentiates of the first session had passed to ordination at the second meeting, and still the number of licentiates was six. Year after year these gifts continued to be bestowed upon the churches. This of itself speaks volumes to the praise of the churches of that section.

Elder J. G. Thomas was the moderator of the first and second sessions. Elder J. S. Allen, who appears as a licentiate in 1855, was in 1856 an ordained preacher, and pastor of the Prospect church, Burleson County. Long and faithfully did he serve that church and others, and has rendered much valuable service as a missionary on that field.

At the fifth session, in 1859, held with Elm Grove church, Williamson County, with Elder B. Carroll moderator, there were seventeen churches, with one thousand and thirty-seven members. In 1857 the name of Elder M. Cole appears among the active pastors in this body. He was a bold and earnest defender of the faith, a useful pastor, and a man of great personal piety. Towards the close of his life he toiled in the midst of much physical suffering, and recently died in the triumphs of faith and love. As so many of our preachers have passed and are passing away, we need more such working and praying associations to ask the Lord of the harvest for more laborers.

In November, 1856, while passing through the territory of this association, on my way from Independence, I was called upon to assist in the ordination of Elder T. M. Anderson. He was by birth a South Carolinian, and in 1849 moved to

Texas and settled in Washington County. His membership was in the Mount Gilead church, and by this body he was licensed to preach in 1850. The church at Rocky, Burleson County, appointed the second day of November, 1856, for his ordination, and the presbytery consisted of Elders David Fisher, R. Howard, J. G. Thomas, John Clabaugh, and Z. N. Morrell.

On our arrival at the place some of the ministers were in favor of proceeding at once with the ordination. The minds of others were not so clear as to the propriety of this course, and insisted that the candidate should preach. We may catechise men, and find out very soon whether or not they are sound in the faith, but we must hear men preach before we can decide with certainty that they are "apt to teach." He readily consented, and preached a short, clear and scriptural sermon. We then proceeded to examine his qualifications for the sacred office, and in a concise manner he sated what he believed, and gave scriptural reasons. All being perfectly satisfied, our hands were laid upon him, while prayer went up to God that he might be a faithful minister of the New Testament. He has been a good and true man, and still toils on in the sacred calling.

One of the members of this presbytery, Elder R. Howard, an earnest preacher at the time of the ordination alluded to, had been but a short time in Texas. He was from Georgia, and being full of the mission spirit, he pressed his way to the frontier, and to the destitute cried, "Behold the Lamb of God!" While a man of ability, as a preacher, that which commends him most is the fact that he sought not ease, but sacrifice for Christ.

A large German population had been steadily flowing into the country and settling principally in the western part of the State. Our hearts yearned for their salvation, and yet every avenue of approach appeared to be obstructed. In the town of Independence, a rude German boy kept a cake and beer stand, selling sometimes that which fires the brain to conceive and execute wicked deeds. In order to pass away dull hours, and at the same time to find something to amuse his active mind, he visited the Baptist meeting in progress, with a view of making sport, having been raised a Catholic. He was anxious to become familiar with the English language, and thought that while amusing himself at the meeting he could be learning the language of the preacher. He could understand but little that the preacher said; but the speaker's earnestness and the deep feeling manifested by the audience were so different from anything he had ever seen or heard before, that strange and unaccountable impressions were made upon his mind.

He continued his attendance at the sanctuary, thinking each time he went he would go no more. When the bell rang again, again he felt impelled to go, and soon found himself wishing he could believe and feel as the members of that church did. He was then under the dealings of the Spirit of God, but comprehended it not. Finally God led him to repentance and faith in the world's Redeemer, and Frank Kiefer was baptized. His avocation was given up, and clear evidence appeared that he was a chosen vessel to bear the name of Jesus among his people.

He was licensed to preach in 1856, and after having spent some time as a beneficiary in Baylor University, was ordained to the ministry in 1858. He could now preach in both languages, and was appointed missionary by the Baptist State Convention. Slowly but steadily God enabled him to gain the attention of the Germans, and a church composed of this people was organized on Mill Creek, in Washington County, known as Ebenezer. According to the minutes of 1871 that church numbers one hundred and thirty-one members, supports a pastor for his whole time, and pays liberally to the mission cause. Another church at Cedar Hill, in the same county, composed of Germans, has been organized, and each body has a good house of worship. Besides these, there are scattering Baptists in many parts of Texas, of German descent, that all ere long we hope be sufficiently numerous to have church organizations and preaching in their own native tongue.

Elder Kiefer no longer labors alone. Elder F. J. Gleiss had long labored as a preacher in Methodist ranks. In the providence of God they were thrown together, and after faithfully comparing their views with the word of God as their guide, Elder Gleiss recognized the Baptists as the people holding the true doctrines and ordinances of the gospel, and was baptized. Since his baptism he has been the pastor of the Ebenezer church, and works with all his might on that field. He is a man of strong mental powers, and displays ability as a pastor second to but few in Texas.

Two years ago Elder F. Heisig came to Texas, leaving the pastorate of a German congregation in London, England, and has since been giving to the German mission the benefits of his talents and energies. As a Baptist, he makes no compromise with error. He is in possession of a clear and discriminating mind, has enjoyed the benefits of a thorough education, and is eminently qualified to oppose those systems of error that have taken such a deep hold upon the German mind. He is under the patronage of the mission board of Union Association, and while he preaches with great power in his native tongue, preaches also acceptably in English.

In addition to this noble trio is Elder C. M. Hornburg, the pastor at Cedar Hill. There are, in addition, several pious young men qualifying themselves to enter this work. Elder Kiefer still toils on as a faithful missionary, beloved by the people, speaking both tongues, and is, in addition to his work as a preacher, scattering a large amount of Baptist literature among the people. He is possessed of a large stock of common sense, and, being blessed with a liberal education, is well adapted to the position he occupies. In view of the rapid emigration from Germany to Texas, the lovers of a pure gospel should lend all the sympathy and aid in their power to these men toiling for the salvation of a people so hard to reach, and who, when brought to the knowledge of the truth, make such earnest workers in the vineyard of our Lord.

CHAPTER XXX.

SKETCHES. — 1857 TO 1867

FROM 1857 till 1861 so many were at work and so much was done, that it is a difficult task to pursue further the plan adopted in the former part of the book. A nobler body of private members and a more earnest and efficient band of preachers have seldom appeared on any field than the Baptists had in Texas during that period. The increase in our numbers since is clear evidence of the truthfulness of the statement.

On the first day of November, 1857, I was called on to part with my last child. A. H. Morrell, of whom notice has been given in the preceding chapters, fell asleep in Christ on that day. Four children were given me, and four were taken away, and their mother; and still God has permitted me to survive them all for fifteen years. My name has since appeared in the reports from the patent office of the United States as having recieved four patents. These rights to inventions were taken out, according to the request made by my son on a dying-bed, and were the results that followed the experiments before alluded to, with ploughs and harrow following after wheels. Misfortune, and the condition of the country in consequence of the war, prevented me from realizing anything from them.

During this same year I was present at the organization of the Austin Association, with the church in the city of Austin. Thirteen churches were represented. Hon. E. D. Townes was the first moderator. Among the many brethren in the ministry, and among the laity, of piety and intelligence, I met Elder A. W. Elledge, according to my recollection, for the first time.

He came to Texas a few years previous to this, and in 1854 and 1855 his name appears as one of the missionaries of the Baptist State Convention. While serving in this capacity the Lord blessed him with much success. In 1857 he was in charge of three churches, Barton's Creek, Walnut Creek, and Bethlehem. His ministry has been confined to middle and western Texas; and wherever he has labored, saints and sinners all remember him; but especially has he made an impression upon that class of men who take pleasure in opposing Baptist principles, and who in their haste are sometimes guilty of misrepresentation. Such men on his field are never allowed to pass without rebuke. Girding himself with truth, and possessed of the same spirit that moved Paul when he "fought with beasts of Ephesus," he shows such men no quarters. While he abstains in his ordinary ministry from all unfriendly attacks on other people, his lion-like boldness is proverbial when the defence of the principles he holds so sacred becomes a necessity.

His mind leads him to do mission work, and in this field his soul is happy, while his masterly native intellect grasps the word of life and points the trembling sinner to Calvary. While serving as missionary in the State of Mississippi, he baptized more than a tousand persons, and his name is sacred to many in Texas, by reason of their submission to baptism at his hands. The last sermon I heard him preach was in August, 1871, at the close of the revival at Eautaw, in Limestone County, from the words, "The harvest is past, the summer is ended, and we are not saved." His clear understanding of the work of God, and his earnest zeal in that effort, revealed the great secret of his power as a preacher. He still lives and labors as missionary in Waco Association.

Representatives from thirteen churches, with a total membership of five hundred and thirty, met with the Mount Zion church, Rusk County, on the thirtieth of October, 1857, and formed the Mount Zion Association. Six ordained and six licentiate ministers were members of this body with Elder J. H. Rowland moderator.

The year 1958 was one of harmonious activity and progress among our people. Five new associations sprung into existence. Four of them — Richland, Leon River, Brazos River, and San Marcos — were west of the Trinity River; while a fifth, Tryon, — named after him who in an early day did so much to give the Baptist cause position in the public mind, — was composed of churches on both sides of the river. The aggregrate number of members that formed these bodies was two thousand five hundred and thirty.

I was present and took part in one of these organizations, Leon River Association. Great destitution had prevailed in the counties of Bell, Williamson and Coryell, until the voices of Elders John Clabaugh and John McClain declared the way of salvation to the people. Both of these were earnest men and men of God. They were well adapted to such a field, and the people heard them gladly. Possessed largely of the spirit of Christ, they were willing to work, and their labor was of that effective kind that moves and induces the multitude to feel that "Jesus of Nazareth passeth by."

As the reader will remember, in an early day I travelled much over that beautiful country that lies between the Colorado and Guadalupe Rivers, south and west of the city of Austin. Dangers then hung around our path at every step, and anxiously did we inquire "how long" until the savage would pass out, and school-houses and church edifices would be erected, whose bells would sound the funeral knell of Mexican and Indian depredations. After light began to dawn upon our dark sky and the campaign of 1842 ended, it only required sixteen years to afford a sufficient population to give the Baptists strength enough in the Colorado Association to spare seventeen churches to form the San Marcos Association in November, 1858. Right nobly have the brethren in that territory done the work assigned them by the Master.

Elder J.V.E. Covey appeared among the ministers and educators on that field. He came to Texas in 1853, and settled east of the Trinity, in the town of Palestine. For about two years he taught a very flourishing school at that point, and in the mean time served as pastor of the church there, and at other points in the same county. A large body of young men received tuition from him while at Palestine, and some of them have made their mark in the learned professions since. His ability to inspire young men with energy and noble aspirations, and at the same time to maintain a firm discipline, is a combination of talent that must always give him success as a teacher. While he was able to do much in the east to encourage the spirit of missions and the cuase of education, his mind led him west, and his name appears among the ministers of the Colorado Association, in 1856. He preached and taught for several years in Lavacca County, and afterwards went to Concrete, in Dewitt County, near the Guadalupe River, where he has for several years conducted one of the best schools in the west. His influence upon the Baptist cause, and over a large number of the rising men and women in Texas, will be felt long after he has passed away.

Casting my eye over the recent records made by Baptists in the west, it affords me much pleasure to see the names of such men as Elders Pinkney Harris and H. M. Burroughs in charge of strong chuches on the Colorado and Guadalupe, where once a few of us worshipped with carnal weapons buckled on, ready for action when the guard, with Texas rifles, gave the signal that the enemy was at hand.

Elder Burroughs, although for several years deeply impressed by the Spirit of God that it was his duty to preach, was naturally timid and troubled with misgivings. The date at which he entered upon an active ministry I cannot give. His name appears in 1866 as pastor at San Marcos and Seguin. Along the valley of the Guadalupe he has done much since to develop personal piety in the churches, and has been effective, by his mild, persuasive address, in winning souls to Christ. His meekness, earnest piety, and plainness of speech constitute the elements of his strength as a preacher.

Elder Harris is the present efficient pastor of the church at Plum Grove, on the Colorado, that was organized thirty-three years ago. Under his ministry the church has been active and greatly strengthened. He is a graduate of Baylor University, having remained for some time under the tuition of R. C. Burleson, during the time he was in charge of the institution. He was ordained while a member of the church at Independence, in 1860. Although he is at times troubled with a little impediment in his speech, when thoroughly aroused his thoughts burn and his words cut. A man of strong impulses, he feels deeply and speaks boldly. Such men rarely fail, if once subdued by the Spirit of God, to make deep impressions upon the public mind. With a strong native intellect, invigorated by much earnest thought, he delights to wrestle with the profound truths of the Bible, and then in his sermons becomes deeply moved as he sees Christians feeding upon the strong meat of the gospel.

Much of his time has been spent in teaching, and in attempting so long to do the work of two men, he is old while he is young. He has succeeded as well as a teacher, and along the valley of the Colorado, from Austin to Lagrange, where most of his preaching has been done, he has made a deep impression by his earnest ministry.

Those who spent the years 1859 and 1860 in Texas will never forget the fearful drought and withering heat of those seasons. The atmosphere, at one time, felt very much as though it had issued from an oven. The corn withered before the time for maturity, and the ear so fondly looked for by the husbandman did not make its appearance on many fields, leaving the laborer dependent on other people for his bread. The cotton bowed its head and lost its power to retain the form that develops the snowy lint from which much of our clothing is manufactured. The Lord did not speak to us with an audible voice, but by the manifestations of his power to give rain or withhold it, he seemed to say, "The heaven that is over they head shall be brass, and the earth that is under thee shall be iron." The lowing herds on our beautiful prairies were, in many instances, driven a great distance to some of our rivers that continued to flow, and many a cow and sheep and horse wandered in search of water, with no herdsman to point out the way, and perished.

We did not then understand it, but to many of our minds since the reason is apparent, that the mildew on the grain, the falling of the cotton leaf, and the fearful indications of famine that made many a stout heart tremble, were merciful dispensations of Providence that prepared us for the darker days of war that were nigh at hand. The political sky was even then already dark, and muttering thunders of the coming storm rolled across the heavens.

Many grew desperate and had not God in all their thoughts, while the more pious sought the Lord by prayer for refreshing from his presence. Showers of grace fell on many localities, from a prayer-answering God, and many churches were greatly strengthened and increased in number.

During this time Elder F. M. Law made his appearance in Texas. I think he came, in 1859, from the State of Alabama. He settled in Washington County, and for some time had the care of the churches at Brenham and Providence. In the spring of 1860 I first met him, and was at once impressed with the fact that he was a man of sound judgment, and an earnest working pastor. An intimate acquaintance with the man and his work since has intensified the impressions I then received.

He settled as pastor at Plantersville, about the beginning of the war, and during the struggle rendered some efficient service as missionary among the soldiers. He labored at Plantersville and Houston until the fall of 1867, and early in 1868 assumed the leadership of the Baptist congregation in the city of Bryan. Being at that time the terminus of the Texas Central Railroad, he had a fine opportunity to develop his ability to rally scattered forces, and build up an interest in the midst of many difficulties. He was, however, the right man in the right place. The church has steadily grown under his ministry, and his faithfulness as pastor has gathered about him and maintained a constantly increasing affection and confidence on the part of his people, until the present time. But few better pastors are found in any country. In person he is of medium height, and with an apparently delicate structure. He is wonderfully gifted with ability to influence and control the actions of men.

In the pulpit he seldom rides in balloons, and has better sense than to attempt to wade in waters where the mightiest men sometimes get strangled. While eminently sound in the great doctrines that have ever distinguished Baptists from other people, his preaching is plain and practical. Strangers are sometimes slightly repulsed when he knits his brow under the pressure of deep and concentrated thought; but his people, who know him well, understand that it is an indication of that deep earnestness that has fired their hearts on so many occasions to labor and sacrifice. Our educational and mission enterprises find in him an earnest advocate.

It was my privilege to be present and take part in the organization of the Waco Association in November, 1860. Representatives from nine churches, with a total membership of five hundred and forty-one, met with the church in the city of Waco. At the first session, the Waco Classical School — since known as Waco University — came under the patronage of this body. The first clerk of the association , Brother J. W. Speight, was, and has continued, president of the board of trustees of this institution.

His indomitable energy put forth in the erection of buildings and providing accommodations for the school has but few parallels. Full of the spirit of enterprise, an ardent friend of education, and in favor of a strict construction of the great constitution that Baptists have sworn to support, he is, and has been, a man of extensive usefulness. He is a lawyer of acknowledged ability, and, as is characteristic of men in that profession, is willing to discuss and ventialte all questions containing vital principles which come before our general convocations. In the midst of these discussions, brethren that don't visit the courts often would imagine he was angry. And many a pious lawyer has suffered in reputation at the hands of his brethren. At the bar, where heated debates are of daily occurrence for weeks, and sometimes months, swords are whetted so sharp that they often cut unawares. In consequence of this fact, brethren should not be unreasonably sensitive towards our friends who practise at the bar, and these brethren, on the other hand, should carefully guard their language in our general meetings; and by these efforts, avoiding sensitiveness on the one hand and undue sharpness on the other, "the unity of the spirit in the bonds of peace" may always be preserved.

It has been my lot to cross swords with brother Speight on more fields than one, but knowing that he was a good and true man, and he knowing at the same time the impulsiveness of my nature, sharp words and remarks that appear to be personal are laid by both of us at the foot of the cross, and affection and confidence burn the warmer upon our hearts.

Intelligent and active brethren, full of zeal and enterprise, have been the moving, leading spirits in this body since its formation. Elder N. W. Crain, who came to Texas at an early day, when quite young, and who had long been in eastern Texas, was one of the ministers present at the first meeting, and was the pastor of the church at Bosque at the time. Only a few years previous to this he entered the ministry and moved into the vicinity of Waco. With a strong mind and a liberal education, backed up by a vigorous constitution, he possessed the elements of usefulness and delighted to labor among the people at large. He is on that field still, and the pastor of the church in East Waco. When I first met him he was a man of large means, financially, and while his steps have been so ordered as to deprive him of a large amount of this world's goods, my last interview with him convinced me that he had grown in grace and knowledge and with the blessing of God was entering a state of more extensive usefulness than he had ever enjoyed in the past.

While this body has estimated highly the cause of education, they have been earnest in their endeavors to advance the mission cause, and have ever manifested a willingness to compensate liberally those who serve them on the mission field. During the past year Elder W. W. Harris served them with a liberal support.

He and Elder Pinkney Harris were ordained on the same day at Independence, in 1860. His name appears in the minutes of the San Marcos Association in 1861, as pastor at Bastrop and Hill's Prairie. He was educated at Baylor University, under the tuition of Elder R. C. Burleson, and has devoted himself exclusively to the ministry, in western and northern Texas. His powers developed rapidly in his early ministry, and the fame of the young man was soon coextensive with the Baptists of Texas. From the very commencement he proved himself a man of genius. His descriptive powers were of high order, and chaste language flowed from his lips without the least apparent effort.

Soon thoughtless brethren joined the name of the great preacher in London to his, and he was spoken of as "Spurgeon Harris." Many brethren, at a distance, only knew him by this appelation. Such things I have ever considered in bad taste and opposed to the spirit of Christianity. While many a young man has been sadly injured by these flatteries, brother Harris pursued the even tenor of his way, and has rendered most valuable and effective service on many fields. The grace of God, be it said to his glory, has been sufficient.

He is a man of eccentricities, — by no means intentionally so, — and right in connection with these deviations from the usual customs of men are sometimes flights of oratory that cause the brain to reel while in pursuit of the speaker's train of burning thoughts, clothed in language that nothing but the most vivid imagination could invent for the occasion.

Under the pressure of great labor, frequently holding meetings of weeks' duration in different localities, he is permaturely growing old. His burning appeals have rung in many ears, and through him, as an instrument in the hands of God, many souls have been turned to righteousness.

Only a few months ago I was permitted to pass along the streets of the beautiful city of Waco, with its thousands of inhabitants, its splendid rows of costly buildings, its magnificent bridge that spans the Brazos River, the large buildings in addition, that furnish the rising generation and their teachers the facilites for receiving and imparting knowledge, and was happy to see that in the midst of all this enterprise a pure gospel was still upheld by an efficient body of Baptists on the same spot where Elder N. T. Byars, so many years ago, in the name of God, organized a feeble church, and where the Waco Association was formed in 1860. Elder H. Carrol, the rising man in the Texas ministry, is a member of that church and association, and is, and has been for some time past, the pastor of the Waco congregation. He, too, is a graduate, having recieved tuition from Elders R. C. and R. B. Burleson, who have educated so many Texas preachers. His ministry has been short, commencing only a few years since, in Burleson County, in the bounds of the Little River Association. He is possessed of an active, inquiring mind, and has enjoyed, in point of education, the best facilites that Texas could afford. God has honored him with much usefulness in a short space of time.

Tall in person, and commanding in manner, he takes bold posititons as a preacher, and is destined by the blessing of God, with patient toil, to do valiant service on the battleground of truth. Long live the young and rising preachers in Texas, after the soldeirs of the old guard sleep in the dust!

The San Antonio Association was organized in 1860, but the minutes of none of its early sessions are before me. This is our extreme western organization.

Dark were the clouds, and fearful as dark, that hung over the churches in 1861, when the sound of war was heard. My heart grows sick at the thought of pursuing my investigations during the four years that followed, and I leave the record for others to make.

In the spring of 1865 the smoke of battle was brushed away, the "confused noise" of war was hushed, and "garments rolled in blood" were lain aside. Having been so stupefied by long and active sympathy with the suffering and the bereaved, some of us could scarcely realize what an ordeal had been passed through. Emerging from the waves that had been sweeping over us, the strongest evidences at once appeared that the Baptists, in their organization, were according to the divine plan. Among them, as a people, there is no great centralized power, in pope, in ruling bishops over large territories, or in councils, that assume the right to make laws and dictate the manner of their execution. Every church is a separate and independent republic, or rather a theocracy, — knowing no head but Christ, and recognizing no laws save those instituted by him. With this fundamental idea, founded upon the word of God, that every church is a sovereign body and independent of every other church, that every ordained minister is a scriptural bishop, and independent of every other bishop, this people has lived, and can live, under all forms of government and pass through any kind of human revolutions, and maintain their unity and organization. With "one Lord, one faith, and one baptism," written as our motto, we may, like the king of Israel, say, that "in the name of our God we will set up our banners" with confidence in the promise made by Chirst, as the "Chief Shepherd" and "Captain of our salvation," who is the head of the true church,and against which, according to his declaration, "the gates of hell shall not prevail."

The churches maintained their organizations well, and many were largely increased numerically. In point of number, in my opinion, we had not lost, but gained, notwithstanding the many that were lost by death in the war. Rallying our forces in our general organizations again, a noble band appeared ready for work, with only here and there a brother, and in a few instances a minister, whose strength had not been sufficient for the temptations and trials incident to that terrible period.

A large number of brethren from other States cast among us, and with these came some excellent ministers. Elder Wm. Carey Crane, a Virgininian by birth and education, and who had labored long and effectively in Georgia, Mississippi and Louisiana, came to Texas in the fall of 1863, in answer to a call from the trustees of Baylor University, and took charge as president of the institution. He remains at that post still, with prospects of future success, and has under his tuition seven promising young ministers preparing themselves for intelligent and active service in the Christian ministry.

As a scholar, Edler Crane has but few equals; and superiors are very scarce. His conversation, his literary address, and his sermons all show that he is not only a profound scholar, but that he has always been a student, and is a student still. His mental discipline has been of the most rigid character. In person he is of medium height, with compact form inclined to corpulency. He has a vigorous constitution, and but few men are able to do the amount of work he does.

When I first saw him I thought his manner was somewhat haughty and stiff. Each time I met him afterwards, I saw more plainly my mistake. I can now say, that a more social, kind and loving spirit it has rarely been my lot in life to meet. His kind consideration and affectionate demeanor towards his brethren in the ministry, who are his inferiors in point of education, I do think worthy of imitation.

His power as a preacher, when fully in the spirit of his great Master, can only be understood and appreciated by those who have heard him when he was moved and stirred by the soul-inspiring, experimental truths of the gospel. Under the pressure of all the duties of a college president, he preaches regularly, and to his ministry devotes much thought. He is doing a noble work for the churches and the cause of education.

Elder J. B. Link, who went out from Missouri with the army and served among the soldiers during the war as a minister, was in Texas at the time of the surrender, looking after the army mission work under the appointment of the Domestic Mission Board. Seeing the vast resources of this great State and the Baptist strength to be developed, he conceived the idea of starting a denominational State paper. A few brethren of intelligence and enterprise encouraged the undertaking, and about the close of 1865 he issued the first number of the Texas "Baptist Herald," from the city of Houston.

He was entirely without means, as was many a man who has served with the army. Traversing a large part of the State, he made many friends to the enterprise that he so fondly and resolutely cherished. Facilities at that time for travelling through the State were by no means good. When he could not conveniently get a horse, he walked from one locality to another, and by his prudent course, sound sense, and indomitable persevance, he convinced the Baptists that he meant work as a means of success. Such men rarely fail to get help if they need it. His case was not an exception. Capital was soon furnished to start the enterprise. His ability as a financier was soon apparent. Economy was most rigidly practised. His dress, as many of us remember, was very plain, consisting for some time in part of that he had worn in the army. All praise to the Texas people. no man has ever come among us since 1835, if he had personal merit, who failed to be appreciated, even if he appeared to be "a poor man in vile raiment." He labored hard, and after his day's work was done his bed at night was a pile of carpenter's shavings in the corner of his office, and his covering his soldier blanket. His very life-blood was thrown into the paper, and as that valuable sheet goes in many instances where the author of this volume never expects the book to be seen, it is not necessary to dwell upon its merit. During these years past the paper has done a great work in developing Baptist principles and in organizing Baptist strength.

CHAPTER XXXI.

THE LAST TRIP OF A PIONEER. — 1863

LOOK on the map of the Western Continent, and among the States of Central America you will find a territory known as British Honduras, nine hundred miles south of New Orleans, bounded on the north by Yucatan, on the west and south by Quatemala, and on the east by the Bay of Honduras. The length of this State is two hundred miles from north to south; its width at the southern extremity thirty miles, and it is sixty miles wide on the north. It has eighteen navigable streams, as far as the tide-water extends, and on some of them vessels may go beyond the tide into the interior of the country. It embraces the 16th and 17th parallels of latitude north of the equator, with a mild and even temperature. In the warmest summer month the thermometer never rises above 88 degrees, and on the coolest winter day never goes below 60 degrees. The people of that country claim the protection of the British flag. As the country is owned by the English government, the laws are written in English, and all the inhabitants speak the English language.

In the spring of 1867 my attention was called to that country, and, after obtaining the facts just recorded, I was impressed with the belief that a large number of people from the United States, especially from the South, would eventually make it their home. Under this impression I sought the Lord in prayer as to my duty in that event. During the fall of the same year my mind was made up to go, "if the Lord will," for the following reasons.

Thirty-two years of labor, with much physical suffering, in Texas, had brought about a state of debility that rendered me entirely unfit for active service as a preacher. In search of a milder climate, I left the snow and ice of Tennessee in the winter of 1835, and sought a home in the wilderness of Texas. The climate suited me, and I was able to prosecute my work as a preacher. Texas northers had grown too severe for me, and I felt a disposition to seek a clime where snow never falls, and where ice and frost are never seen. Honduras presented me this inducement, and the improvement of my health was a leading consideration in the change.

Information soon reached me that a large number of southern families intended and were making arrangements to emigrate to Honduras. The heads of these families were mostly men of intelligence and enterprise, whose estates had been destroyed during the late war, and who wished to enter a new country with a view of recuperating their wasted fortunes. Among these I was confident would be Baptists, who would be greatly in need of a ministry and church organization. The Baptists of Texas had been organized, and as the principal part of my ministry had been given to the destitute, I still felt inclined, in the language of Paul, to preach the gospel in the regions beyond, and not to boast in another man's line of things made ready to hand.

A celebrated character in the Revolutionary war was Captain John Hunter, from North Carolina. He afterwards turned Baptist preacher, and was strangely impressed by a dream connected with my future history. He was a man of great eccentricity, and made strong impressions upon the minds and hearts of almost all whom he met. He was an old man in 1823, when I was quite a young preacher. So strongly was he impressed with his dream that he came some distance to see me in Tennessee, and to communicate his impressions.

He dreamed, and stated as his conviction that it would come to pass, that towards the close of an eventful life I would cross the great waters and preach the gospel in a strange land, and after remaining there for a while, would return to the United States, where my light would gradually go out. I often afterwards thought of the earnestness with which the old preacher rehearsed it, but had no confidence in the reality of his prediction. When my mind was made up to go to Honduras, I remembered in tears the counsel of the old preacher, long passed to his reward, in view of the conflicts and trials that he felt confident awaited me towards the close of my life.

The fearful scourge of yellow fever that visited our southern cities in the fall of 1867 prevented me from leaving Texas as early in the season as was desirable. In December, reposing my confidence in the God of the whole earth, I left Texas, and, after a short delay in the city of New Orleans, took a berth in a vessel that rode the high seas between New Orleans and the city of Belize. On board I found nearly one hundred passengers going to the new country, and among them about a dozen Baptists.

Just before the ship went to sea I felt impressed to preach to the passengers, and although unable to stand without a support, I stood with one hand grasping a post in the cabin, and from the following text preached a short sermon to an attentive audience: —

Hebrews xi. 14: "For they that say such things declare plainly that they seek a country." Some of the advantages of the country to which we had just taken passage were used to illustrate the beauties and excellences of the land that partirachs and prophets sought by faith.

In the month of January, 1868, we were landed in the city of Belize, with a population of fifteen or twenty thousand souls. From June until the first of March is considered the "rainy season," and a large number of the inhabitants of the country who spend the "dry season" in getting out timbers, and in planting little crops of various kinds, were at this time in the city. During the dry season the population is not more than five or six thousand.

The population of the whole country known as British Honduras was, at that time, supposed to be about thirty thousand. Of this number there were about one thousand English, one thousand Americans, and the rest natives, — negroes, Spaniards and Indians. A large majority of the inhabitants are negroes. Among these may be observed the same shades of color that mark the colored people of our southern States. These people are, as a class, superior to the negroes of America, both in physical structure and intellectual capacity. A system of education prevails that enables nearly all of them to read and write, and, in many instances, they give evidence of strong mental powers.

In no country that I have ever visited are the laws so rigidly executed as in Honduras. The natural consequence is the cultivation of a high grade of personal honesty among the natives. Often was I reminded of the great contrast, in this respect, between the United States and this part of Central America. Violations of law are rigidly ferreted out, and a just punishment inflicted upon the violator. Personal honesty is thus cultivated to such an extent that the emigrants soon felt inspired with confidence that the natives would neither steal nor defraud. In one instance I was followed by a colored boy for some distance, who returned to me six and a quarter cents that I left at the store of a negro merchant. After I left he discovered the mistake I had made in paying for what I had purchased. In another instance, one of the passengers who went over when I did, being addicted to strong drink, concluded that he must have a regular "spree," and, during a fit of intoxication, left sixty dollars in silver on the counter of a tailor. He and his friends at once gave up all hope of seeing the money any more, as he did not even remember where he left it. Passing the same door, some time afterwards, he was recognized, and the money returned without even so much as having been unrolled. Incidents of a similar character frequently occurred during our stay in the country.

We found the Sabbath day rigidly observed by all the people. No doors were open, and no business of any character transacted that could possibly be avoided. In the city we found two Baptist churches, one Wesleyan Methodist, one Presbyterian, one Episcopal, and one Catholic. Whatever form of religion the people observed, they followed up regularly and earnestly.

Here I met the venerable Baptist missionary, Alexander Henderson. He was about sixty-five years of age, and had been in Honduras thirty-five years. He was a native of Scotland, and in early life entered the ministry as a Presbyterian. He was educated in his native land, and afterwards preached in France. While there he formed the acquaintance of a Baptist preacher, and an intimacy grew up between them that led to a free discussion of their differences in doctrine and church polity. He was a ripe scholar, and, basing his investigations upon the Scriptures as they were originally given, decided that the Baptists held the ordinances as they were given by Christ, and that they observed a scriptural church government. He at once became a Baptist, and tendered his services to the English Baptist Missionary Society, expressing a willingness to serve as a missionary in a foreign land. Under the auspices of this society he entered British Honduras, and had given his manhood and strength to Christ on that field. Brother Henderson held and rigidly defended the views of communion as practised by the Baptists of the United States.

Some differences sprung up between him and another missionary, appointed to this field by the same society in later years, on the communion question, and rather than serve under the patronage of the society who would encourage open communion on the field where he had toiled so long, he resigned his commission, and was serving as a missionary at his own charges. He is eminently sound in the faith of the gospel, and hesitates not to declare, on all proper occasions, the principles that have ever distinguished Baptists from other people. His congregation in Belize supported him liberally, according to their means; but it was necessary for him to spend a small portion of his time in secular matters to gain a living for his family.

As a scholar he was held in high esteem, both in England and in Honduras, and he spent a large portion of his time in translating the word of God into two foreign tongues. The most rigid system in his work and economy in his time were maintained. Not an idle hour was spent, and yet just so much time was given to translation, to his work as a pastor, and to his little business, upon which he was compelled to rely in part for his living. In this way a large amount of work was accomplished, for a man of his years. Many of our ministers in this country work as hard as brother Henderson, and yet accomplish scarcely half as much for want of system.

System is the order of the day in Honduras, among all classes of business men. Stores all open at daylight. When the city clock strikes nine, every door, by common consent, is closed till ten. This is the hour for breakfast. When the clock strikes ten, every business house opens instantly. At four in the afternoon every house is closed for the day, and until nine o'clock at night the people all devote themselves to leisure and recreation. Good bargains and money-getting will not induce them to violate these rules. In consequence of these habits of regularity the people are healthy and happy. American business men work all day, and, in many instances, a large portion of the night, and then are dissatisfied because tired, exhausted nature can do no more. No such disposition is manifested among that people with whom my lot was cast for many months.

In keeping with the other regularities of the country, the old Baptist preacher toils on and wields a healthy and happy influence over the inhabitants of his field. The writers is largely indebted to him for many kindnesses shown a stranger in a strange land, suffering greatly from bodily infirmities. Eternity along can reveal the amount of good brother Henderson has accomplished among that people.

In that climate the heat is very seldom the least oppressive, and there is never need of fire except for cooking purposes. Uniform good health prevails among the people, and many of them live to a great age for this period of human history.

In consequence of the infirmities which I carried with me, I was seldom able to visit the interior, for want of proper conveyances. The following are some of the observations made as opportunities were given me to penetrate the country.

The soil is of three kinds. The larger portion of it is chocolate, and very productive. Some of the land is black and sticky, very much resembling the hog-wallow prairies of Texas. Besides, there is a large amount of gray sandy land, easily cultivated. Representatives from six Southern States went over with me, and our decision was, that the lands up to the foot of the mountains were equal to any in the Red River or Brazos valleys for agricultural purposes.

None of the common implements of husbandry had been carried there, and the crops were planted and cultivated under very great disadvantages. The undergrowth in the land that knows no frost, of course, is very dense. This is cut down at the beginning of the dry season, and at the proper time is set on fire. The earth, after this burning, is left like a plant bed. As soon as the rain begins to fall, a small hole is made in the ground with a stick, the seed deposited, and covered with the foot. Nearly all of the seeds planted in our southern States are grown there, and the tropical fruits besides. The soil and climate are admirably adapted to sugar. Cotton is not a very reliable crop. The lands I saw in cultivation were not ploughed; and with the bushes and weeds cut down, as the only work done, good crops of corn, potatoes and other productsions common to this climate were returned to the husbandmen.

In that land the tropical fruits are found in great abundance. The banana and plantain have been described by others, and the fruit in large quantities is transported to our shores. But the difference between these fruits and the orange of Honduras, plucked full ripe from the tree, and those sent us across the waters, is about as great as the difference between a fresh ripe lemon or peach, and one a week old after the stem is broken. The custard apple is one of the delicacies of the country, and will not bear transportation. It grows on a small tree three or four inches in diameter, and some ten or twelve feet high. The tree bears mature fruit at the age of three years, very much resembling the egg plant. The apple is about four inches in diameter, and when fully ripe is in appearance a dark mixture of red and purple. The skin is as thin as one of our common apples, and very tender. The fruit is very soft, yielding easily to the pressure of the spoon, and combines both sweetness and acidity, that render the taste of the most delicious character. No lady in America is skillful enough in the art of custard-making to produce a dish superior to the custard apple, fully ripe, and fresh from the tree.

One of the most interesting productions of Honduras, to the student of the Bible, is the palm-tree, spoken of by David in the ninety-second Psalm, and in other parts of the book of God. Two varieties of this remarkable tree arrested my attention, and engaged much thought. One is known among the natives as the "cabbage-tree." The bud is very much in appearance and taste the same as our common cabbage, and is prepared for the table in the same way. This is seldom used, as it destroys the tree and requires some labor to procure it.

By far the most interesting species of the palm-tree is that which bears the cocoanut, which is brought to this country, and with which fruit almost every youth is familiar. Standing by this majestic tree, I felt that I was in ultimate association with the thoughts of David in his touching allusion referred to. The history of the tree is as follows: It springs from the cocoanut. This nut may be laid in any place, be it ever so dry, and the milk that it contains possesses sufficient nutriment to cause the sprout to burst the shell and shoot out the single stalk of the coming tree. Not a single root has yet made its appearance, and in this condition they are sometimes carried a hundred miles and planted out in a hole that just receives the nut. It grows successfully only near salt water. Being carefully covered, it begins to put forth very small, fibrous roots, that take strong hold upon the earth, while the tree rises month after month all the year round to a height of one hundred feet. At the age of five years, it has reached the height of ten or twelve feet, and begins to bloom right on top. Eight or ten blooms burst out in the form of a circle, and present a bright golden appearance. In one month these blooms all disappear, and in the place of each one is a small cocoanut, one inch in diameter. These stems that hold the nut right on the end are from one to two feet long, and just beneath is a circular row of limbs from ten to twelve feet long, full of leaves five or six feet long and six inches wide. The limbs are strong and flat on the top, and the leaves are cupped in the form of a gutter. As the cocoanut increases in size and weight, it rests upon these limbs and leaves, causing them to hang over in the form of an umbrella. While this is going on, the tree mounts higher and forms another circle of limbs and fruit above, and every month in every year new fruit forms in the top, while the old fruit at the bottom is dropping off. No doubt the revelator remembered this, when he was instructed to write of "the tree of life which bare twelve manner of fruits, and yielded her fruit every month."

The limbs have answered their purpose when the fruit resting on them is ripe, and according to the arrangement of the God of nature they are disengaged and fall to the ground, leaving a scar on the tree as the only damage. These trees never quit bearing fruit as long as they live, and hence the psalmist says, "The righteous shall flourish like the plam-tree. They shall bring forth fruit in old age.'' An English gentleman pointed out to me some trees that had regularly yielded their increase for seventy-five years.

No limbs ever remain on the tree except what are necessary to bear up the fruit, and by this I was forcibly reminded of the necessity of laying aside all the encumbrances of this life that will not help us bear ''fruit unto holiness.'' The scars on the tree, every one of which is plainly visible while it lives, illustrate to the thinking mind the scars the Christian will wear when at the end of life's conflicts he receives the palm of victory.

There are no large roots beneath this tree, but a million of small fibrous roots permeate the earth in every direction for some distance, that serve the double purpose of strongly supporting the tree against the winds, and of appropriating the growth of the tree and the maturing of its fruit all the productive properties of the soil near by. Nothing grows near a large palm-tree. So the Christian's hold upon the earth is by a multitude of small acts and silent influences, and these, like the vast number of roots that support the palm-tree, should be so multiplied as to lay every influence we can command under contribution to the glory of God.

The country is very heavily timbered, and the trees are generally large and very tall. Mahogany, logwood, rosewood and cedar, of the best quality, are found in great abundance. Long ranges of mountains appear, some of them of great height, and from these come rolling down into the plains beneath some of the most beautiful streams of water that ever cheered a weary traveller. The water is pure and clear, and in many instances the streams are strong enough to furnish propelling power for machinery, either to grind food or cut the timber that abounds on every hand. Fishes of the best quality abound in all these streams.

With all these advantages, many a youth who reads these lines will feel that it is a better land than Texas. In many respects it is, and yet the disadvantages are not a few. Cut off at present by the mighty deep from England and North America, many years must pass ere the country can be settled by a sufficient number of our people to make a home in that land desirable. As the order of the day is to belt continents and islands with parallel iron tracks, thus sending travellers by steam on rapid wheels, the time may come when a sufficient number of Americans can enter Honduras to make society, in point of churches, schools and agricultural appliances, what it is not yet, and what it is not likely to be, so long as families must cross the sea to reach it.

Although the people are generally healthy, yet there are diseases peculiar to the climate and locality of a fearful character. The greatest care must be exercised by new comers to avoid them. As in all southern countries, fevers of a malignant type, and severe disorders of the liver, may be expected, in case of exposure. However, in the city of Belize the yellow fever has never prevailed as an epidemic, although carried there on different occasions. The diseases to be dreaded are dropsy and rheumatism. Among the laboring classes these diseases prevail to a considerable extent. Exposure in the rainy season induces these troubles, and patients frequently linger for years, amid great suffering, and finally die. The mildness of the temperature and the fresh sea breeze greatly strengthened and improved my feeble lungs, but dropsy seized upon me in a severe form, and, had I remained, would long before this have terminated my life. With all this, a vigorous constitution, with prudence, in my judgment, has a better prospect for long life in Honduras than any part of the world I ever visited. The temptation to exposure is very great, as it is never disagreeable to wade in the water, or expose the person to the falling rain.

Two crops of grain and many other things may be raised every year; but the great abundance of rain that floods the earth during the long rainy season will, in my opinion, always prevent that thorough cultivation of the soil that is necessary, in order to keep a farm in proper condition for a succession of years. With the present system of cultivation, grass and weeds, in the absence of frost, soon take possession of the ground to such an extent that the field is given up and another one cleared.

Most persons would suppose that such a country would be infested with snakes of the most poisonous character; but in that land they are very scarce. Insects of the most troublesome kind annoy the American greatly, while he eats the custard apple and other delicious fruits, and refreshes himself in the soft, balmy atmosphere. Promiment among these is the sand-fly, with which the people of our country are well acquainted; but my experience is that these flies are at times more numerous in Honduras, and *better biters,* by far, than any I ever met elsewhere.

Next to these is the chegre, a malicious insect, a little larger than the red bugs that used to annoy me greatly when a boy, after playing over mossy logs. This insect does not operate alone, but, in some way I cannot explain, possesses the power of multiplying into a large number. A company of these made, on one occasion, quite a fierece attack on my left heel, and before they were removed made a round hole in the flesh large enough to receive a pea. An application of turpentine soon relieved me.

A fly, with a short body and large wings, about double the size of a house-fly, infests certain localities among the brush, and deposits a worm on the human flesh every opportunity; that is known as the "beef worm," and it immediately bores into the flesh. It gives no pain while entering, and is usually discovered by the appearance of bloody water oozing out where it entered. A Texan at once thinks, from this description, of a "screw worm," but it rather resembles in its operations the wolf in a cow. As it increases in size, the most uneasy and unpleasant sensations are experienced. An application of tobacco will usually drive them to the surface, and instead of leaving the part inflamed, as would be expected, it heals immediately. Great difficulty is sometimes experienced in getting them out, and in some instances the knife must be resorted to. One of my little grandsons brought some of them in his feet to Texas. No danger whatever, that I know of, attends this insect, as it seems entirely free from poison.

The bottle fly very much resembles the buffalo gnat of Texas and other southern States, and during one month in the year is very numerous and troublesome.

Among other pests is the vampire bat, that very much resembles in appearance the common leather-winged bat of North America, and is about one-third larger. This bat possesses a bill of such a delicate character, and is so skilful in the art of surgery, that it can pierce a vein in the human system and perfectly satisfy its appetite with human blood without waking the patient. While doing this, it gently fans the sleeper with its broad wings, and induces, by the pleasant operation, if possible, a sounder sleep. No harm is done except the loss of blood. I was bled, on one occasion, very freely. This can easily be avoided by using a common mosquito bar. They are a great pest among fowls, and frequently bleed them almost to death.

Although the country has been settled along the coast for a hundred years by citizens under the British flag, no one of them, so far as I could find, could give satisfactory information relative to the mountains and jungles far back in the interior of the territory, that only covers an area of about two hundred by forty miles. There are but few horses there, and on these it is impossible to ascend the steep mountains and penetrate the dense thickets in the valleys between. Most of the wild beasts that live in tropical regions are found in Honduras, and birds and fowls of beauty and interest arrest the attention of the stranger. It is the land of the monkey and parrot. Large bodies of monkeys live in the tall, dense forests, and by their appearance and action, remind us of humanity on a small scale.

Hoping that this brief outline of what I saw and heard of the character of the people, soil and productions of Honduras, may be of some interest to the inquiring mind, I close this chapter by giving a brief statement of facts relative to Baptists in this country.

Brother Henderson, it will be remembered, had been there thirty-five years, and during this time had baptized quite a number of the people. In 1868 the Baptists numbered about three hundred, and, with the exception of a small party, were well organized and active in the prosecution of their work. Four native preachers shared in the labors of the old pastor, and were rendering valuable service. During my stay among them I preached as much as my infirmities would allow, and aided in the organization of one church with about twenty members, and assisted in ordaining one native preacher.

CHAPTER XXXII.

THE CONCLUSION.

A contrast. — Texas as it was in 1835, and Texas as it is in 1872.

IN March, 1869, after spending more than a year in a foreign land, my feet again pressed the soil of Texas. Severe bodily afflictions have been my lot since. During most of this time I have been confined to my room, and yet my mind has been active reviewing the past.

Young men look towards the future, and anticipate grand achievements of an earthly character. Old men, conscious that the sun of life is setting, love to review the past. Following a natural impulse on my part, and with a view of imparting information, the following facts in a condensed form are given: —

Texas has an area about twice as large as Great Britain and Ireland, and is as large as New York, Pennsylvania, and all the New England States.

In 1835 no census had been taken; but from the best information at my command, having travelled over the entire territory inhabited by our people, during that and the follwing year, my conviction was, and still is, that there were about fifteen thousand English-speaking people in the country. Fifteen years later, in 1850, the population was 212,599. In 1860 it was 604,215. In 1870 it was 818,579. Now, in 1872, it is largely over 1,000,000.

Long droughts in early days greatly afflicted us. Now, it rains more in one year than it did then in three. Many have speculated about this; but I account for it strictly upon the principles of providence. The prophet Amos reveals the mystery. Speaking of God, he says, "He calleth for the waters of the sea, and poureth them out on the face of the earth."

In 1835, there was only one newspaper in Texas, the "Texas Telegraph." In 1872, there are one hundred and nine.

Then, there were very few post-offices. Now there are five hundred and thirty-six, and telegraph wires bind nearly all the important towns together.

Then, there were but few plain wagon-roads. Now, there are nearly a thousand miles of railway in running order, and the work goes rapidly forward.

We then had no facilities for the education of our children. Now, there are as many children in school, in proportion to our population, as in any other State in the Union. Nearly all the various religious denominations of professed Christians, in the State, have their schools for the education of the rising generation.

In 1835, there was one Baptist church in Texas, and less than fifty Baptists. Previous to 1865, the white and colored Baptists worshipped as members of the same organizations, and were happy in their church relations. Now the organizations are separate, and the above estimate includes all.

In 1835, there was one missionary Baptist preacher. Now, there are four hundred and fifty white, and over a hundred colored preachers.

Then, there was no association. Now, there are thirty-five, and two general organizations.

Fifty years ago to-day, the joys of faith in Christ were, for the first time, revealed to me. I lay down my pen with a heart full of gratitude, not only because "our salvation is nearer than when we believed," but because of the display of God's mercy and truth in this great State, feeling confident that Texas and the cause I love so much will have a bright future.

CHAPTER XXXIII.

TEN YEARS OF RAPID PROGRESS.

TEN years ago I laid down my pen, expecting never to writer another page for publication. Three-score years and ten I then had passed, and more than half of that time in the midst of disease and suffering that would have brought an ordinary mortal to his grave. I can testify from experience to the truthfulness of scripture statement: "The days of our years are threescore and ten; and if by reason of strength they be four-score years, yet it is their strength labor and sorrow; for it is soon cut off, and we fly away."

My sufferings for ten years past beggar all description. Seldom able to preach, and yet I linger on the shores of time. Why, no one but the Infinite One can answer.

Thankful that my mind has been preserved in the midst of age and suffering, that friends and brethren, young and strong, have stood by me in my debility, and that my hope in Jesus is clear and strong, with the same one to help me who rendered service in preparing the former editions for the press, I will add another chapter concerning things that have occurred since 1872.

No ten years have brought about so great results in Texas as the last ten. The iron track reaches from our city on the gulf of Mexico to the Indian Territory, and has connected our eastern border with the Rio Grande on the west. Mexico and the cities of the east and north are bound to almost every important town in Texas. People of all climes have poured in by thousands; and where ten years ago there would be no settlements at all, or a sparse population, the country now is filled with thrifty people, and towns along our long lines of railroads have grown with corresponding rapidity.

During these ten years Baptist missionaries, colporteurs, evangelists and Sunday-school men have held the fields already occupied, and have vigorously entered upon their work in almost all the new communities that have been formed.

All my grandest expectations in early times have been surpassed. Texas, from east to west, as well as from north to south, has been belted with double iron rails; and now two gates to Mexico stand wide open, and at Laredo and El Paso Texas missionaries stand — born, regenerated and educated on Texas soil.

In addition to these two sons of Texas, who as missionaries of the Cross stand at these open doors, one of our best Texas preachers is preparing to penetrate the very heart of our sister republic, thus binding together Texas and Mexican missions, as Gould, Huntington and others are binding the two governments together with bands of iron. Having with my brethren in early days suffered at the hands of Mexican tyranny, and felt the grinding power of priestcraft at the hands of Roman Catholicism, it affords me indescribable joy to be permitted to bear an humble part in thus helping to sow the seed of "truth as it is in Jesus," not only at the door, but in the very heart of a nation hitherto bound in chains of slavery at the feet of a man who claims to be God's Viceregent on earth.

When I ceased writing in 1872, the indomintable J. W. D. Creath was in charge of the mission interests of the "Texas Baptist State Convention." He was then far advanced in age, but still possessed much of the ardor of youth and steadfastness of manhood. He visited annually, and often semi-annually, a large part of our territory, stirring up Associations and soliciting money for State missions. Having had his mind for some years on San Antonio as the most important field in Texas for Baptists to occupy, he concentrated his energies in gathering means with which to build a Baptist house of worship in that mighty Catholic city. With what little aid the few Baptists and their friends in San Antonio could give, brother Creath raised some twelve thousand dollars, and lived to see a Baptist church assembled there in a good house of their own, and the pastor at the same time occupying a parsonage owned by the church. As soon as the two buildings could be occupied, Elder W. H. Dodson, then pastor of the church at Calvert, was selected as the missionary pator.

Elder Dodson was born February 4th, 1842, in Choctaw county, Mississippi, and was baptized December 27th, 1857. He was ordained in 1867, and spent about three years in Mississippi in pastoral and mission work. He came to Texas in 1870, and took charge of the church at Plantersville, in Grimes county. During the three years he served as pastor of this church there were many additions, and by his zeal and earnest labor a good house of worship was built. From Plantersville he was called to Calvert, and remained as pastor four years. There he was again blessed in his work. A strong church was built up and a good house of worship. Entering the city of San Antonio five years ago, the church numbered ten, with their house of worship yet unfinished. He threw his whole soul, body, mind and money into the work before him, and, as might have been expected, he succeeded. The church now numbers one hundred and sixty-eight, and the same spirit of missions that fired the heart of J. W. D. Creath fires this body. Their energies under their pastor are going out for Mexican missions; and we can well rejoice that the crowning work of brother Creath, ere he passed away, had the mark of permanent success written upon it.

Leaving Brenham in 1873, where I had spent about two years with the family of M. V. Smith, I traveled as much as I possibly could, visiting my former fields of labor, rendering what service I was able in encouraging and strengthening the churches. In 1877, while visiting Mineral Springs in Caldwell county, I was invited by brother R. J. Sledge, of Hays county, to pay him a visit and spend some time with his family. I had known his wife from infancy, having spent much time with her brother, Deacon Terrell Jackson, of Washington county, both before and after she was born. I had formerly been urged by the brethren, members of Martindale church on the San Marcus river, to make that community my home while I should live. Though I felt inclined to accept this kind offer, I have realized since that "it is not in man to direct his steps." Being urged by brother Sledge and his noble wife to protract my visits, their house has become my home, and for the past five years I have received many kindnesses at their hands. May the Lord reward them.

Though brother Sledge is only about forty years of age, by his industry, enterprise and good judgment, under the providence of God, he has accumulated quite a handsome fortune, and has a mind to use it for the good of humanity and the glory of God. He loves Texas and loves the Baptist cause, and appeals for Texas missions are not made to him in vain. May the Lord spare him long, and increase his love and zeal for the cause of the Christ who redeemed him.

I cannot close these personal references without expressing my heartfelt gratitude to the brethren at Martindale, San Marcus, Prairie Lea, Mineral Springs and Luling for their many, many acts of kindness to a poor way-worn soldier, battling with the infirmities of life between the ages of three-score years and ten and four-score years. I needed their sympathy and help, and received both as often as my necessities pressed upon me. It was all done without solicitation. I have been blessed in receiving, and yet my Master has said "it is more blessed to give than to receive."

I had lived west of the Colorado about two years in the early history of the Republic, but was driven back by the Mexicans in 1842. Except one visit west to aid in the organization of the Colorado Association, in 1847, I had not traveled and labored west of the Colorado river since the annexation of Texas to the United States. The joy of my soul was inexpressible as I gazed upon the vast prairies where the savage and hostile Mexican, the common enemies of my people, once made our lives almost intolerable. Waving fields of corn, small grain and cotton told of civilization, while school-houses and houses of worship told that education and religion occupied much of the thoughts of the inhabitants.

I asked myself the reason of all this, and the answer was at hand. All those who are accumstomed to reasoning from cause to effect, and from effect to cause, must know that the larger part of all these grand results had their foundation in Christian missions. Since this field for the past few years has been considered self-sustaining, and our main mission work has been pressed farther west and south, under the ministry of a few earnest and consecrated pastors, a great change has been wrought under my eye during the past five years. Elder Isaac Sellars, by profession a lawyer, formerly living at Lagrange, has been put into the ministry, and is now and has been for the past three years pastor at Luling. He is earnest and active, a good pastor, with the elements of success as a revivalist. During his short ministry many by him have been called to repentance, and with the blessing of God he will in future be a power for good either as pastor or missionary.

He is a native Texan — was baptized by Elder P. B. Chandler in July, 1859. He went to school seven months after he was married, studied law and practiced nine years. He is six feet high, black hair, a keen, penetrating eye, good delivery, full of zeal, and impresses his audience that he is powerfully in earnest. His sermons abound in good illustrations, and, best of all, he succeeds in winning souls.

Near Moulton, in Lavaca county, lives Elder A. S. Bunting. He was born in North Carolina in 1844, brought to Texas in 1846, educated at Gonzales, baptized in 1863, and ordained as a minister at the age of 31 years. He has never changed his location since he became a preacher. Being a Simon-pure Scotchman by descent, he is ardent and enthusiastic in whatever he undertakes.

When I first met him he was in a buggy he had made himself. Prior to his call to the ministry he was a carriage maker. The speed with which he traveled in that buggy has always associated him in my mind with Jehu. A good worker in the Sunday-school and temperance cause, with a clear head and warm heart, he preaches with power and demonstration of the Spirit of God. He is not of that class of men who have to go where they are not known to find fields of labor. The demands around him are more than he can supply.

About four years ago, while visiting the church at Martindale, I first met S. C. Harl. I heard him pray, and was deeply impressed with the fact that the Spirit of God was upon him in power, and that all his energies ought to be given to the ministry. I afterwards aided in his ordination, by the request of this same church. He was a man of limited education, but craved knowledge. Notwithstanding he was a husband and the father of four children, he was willing to enter school with his children; and knowing from sad experience the disadvantages of a minister without literary training, I got his consent to enter school as a student at Luling, and used my influence among the brethren to raise the means necesary. His wife was equal to the emergency; and while husband and children were together in school she took boarders, and did all she could to avoid dependence on the brethren. Without a murmur she has met the labor and hardships of a preacher's wife, and esteems it a privilege to be a co-worker with one whom God has called to preach. His progress in school was good for a year, and will enhance his usefulness for life. My judgment is that, as a rule, men had better continue in the ministry right where God calls them, and study and preach as this brother has done.

Elder Harl thinks for himself — seldom, if ever, repeats sermons — rarely preaches over thirty minutes, and seems to know when he is done. He loves the Bible. He has succeeded. With such a wife to aid him, he will be much to blame if his life and health are preservwed and he does not accomplish much good in future.

It is a standing reproach to the Baptists of Hays county that they have failed to use the means God has graciously given them in building houses of worship. Baptists are shorn of half their strength when they are compelled to worship in union houses or houses belonging to other denominations, either in town or country localities. There is not to-day a finished house of worship owned by Baptists in Hays county. This accounts in a large measure for the fact that other denominations have outstripped us, and for the present weakness of a people who hold doctrines that are peculiar and specially adapted to develop strength and steadfastness in Christian character. To develop strong churches, maintain Sunday-schools and prayer-meetings, and properly impress our doctrines upon the people, we must have houses of our own. In the old town of San Marcus there has long been an organization of Baptists, and with proper effort a house might have been built years ago. Dwelling in houses belonging to other people, the church has had a name and but little else.

Eighteen months ago the Lord sent Elder B. Hatcher to the county town of Hays county, who, though young in years and in the ministry, took in the Baptist situation at once, and wisely undertook the building of a house for the San Marcus Baptist church.

I am by this brother requested to attend the dedication services at the opening of this house of worship, on the third Sunday of June, 1882, at which time I hope to be present. Baptists of any country ought to rejoice when a preacher of ability enters their territory with Baptist meeting-houses on the brain.

Elder Hatcher, though young, is making a strong impression for good upon the people of Hays county. He is pastor at San Marcus, Martindale and Kyle. Physically he does not appear to be strong, and I fear is performing more labor than his constitution will long bear. He is tall and dignified in the presence of an audience; his propositions are clearly stated and logically enforced. If his health is cared for properly, we have a right to expect some good work from him in the future.

"The workers' Institute" of San Marcos Association has rendered some good service in developing the mission and Sunday-school work in Hays, Guadalupe and Bastrop counties. Prominent in this band of workers I have always met Elder T.J. Dodson, now pastor of Seguin. He is by birth a Mississippian, was baptized by Elder James H. Nelson, and is now 28 years of age. He came to Texas in 1869, and afterwards graduated at Independence, Texas. He was ordained in July, 1876, and was called to the pastoral care of the church at Seguin previous to his graduation. But few men hold their first pastorates as he has done. Beloved by his people, he still labors successfully where he began at the close of his college life.

With Judge Douglas, of Seguin, the Moderator of San Marcus Association and President of the Institute, and such brethren as Elders Williams, Fonts, Sublet, Lewis, Porter and others, to strike for progress and truth, the outlook for the Baptist cause animates my soul.

It is only eight miles south-east of brother Sledge's house, my present home, to the place where the battle of Plum Creek was fought in 1840. Four Baptist preachers were engaged in that fight with Jack Hays, Ben and Henry McCulloch and others, who aided in laying the foundations of good government. The Indians were routed and driven with great loss right over the ground on which I now live, and it is a gratifying thought that eighteen Baptist churches are now at work within a morning's drive of that battle ground.

It has not been my privilege for the past ten years to visit the churches in Eastern and Northern Texas; but the mission spirit has been at work there, and other general organizations have sprung up, as the judgement of the brethren in these localities has deemed advisable. Ten years ago the Texas Baptist Convention and General Association of Texas were the only general bodies. Now we have the Eastern Texas Conference, the North Missionary Convention and the Texas Central Convention. The vastness of our territory in the minds of thoughtful brethren make these organizations a necessity, and each and all are pressing on the mission work in their own chosen fields.

It is with pleasure that I record the fact, as I turn from this subject, that under the appointment of the Texas Baptist State Conventin there are three young, active preachers, well known to the writer, all raised and educated in Texas, and occupying prominent outposts on our frontier, south and west. I refer to James M Carroll, John D. Wright and Geo. W. Baines.

As early as 1838, forty-four years ago, I visited the old town of Corpus Christi, on our southern coast. Near that old town, that has since been a stronghold for Satan, I and others came near losing our lives at the hands of Indians. For the past three years Elder J.M. Carroll has stood as our missionary in the face of much combined opposition.

He is the son of a deceased Baptist minister, and brother of Elder B.H. Carroll, of Waco. He has been reared in Texas, is a graduate of Baylor University, and a strong preacher.

At Laredo is Elder John D. Wright, a nephew of Elder S.G. O'Brien, long gone to his reward. Reared by a widow in humble circumstances in the town of Brenham, educated principally in a printing office, he has for the past two or three years done a grand work between Brenham and Austin. He has only recently been called to Laredo, but is laying fast hold of the people by whom he is surrounded.

The last appointment made of which I have any knowledge is Elder G.W. Baines, jr., to El Paso. He is the son of Elder G.W. Baines, formerly Editor of "The Texas Baptist," and now pastor at Salado, Texas. He has been an earnest and successful Sunday-school and temperance worker, and will need all the experience gained from these labors to fit him for the important position he has assumed.

At this grand, coming center of influence in the mercantile world some brethren are proposing to build a house of worship, and call it the "Morrell Memorial." I do earnestly beg that my name may never be used in this connection while I am living, nor when I am dead. By all means build the house, and build a good one. Then let the organization be named The Church at El Paso. This soulds more like Paul teaches, when he writes "The Church at Corinth." than "Morrell Memorial."

At the last meeting of the Texas Baptist State Convention a proposition came from the Home Mission Society of New York, to the effect that for every thousand dollars Texas would raise for missions they would raise a like amount and expend it on Texas missions. With this aid Elder O.C. Pope is now canvassing our State, raising money and securing missionaries. Thus the same organization that aided in an early day in the support of Elders Tryon, Hucking and Taliafero, is still anxious to lend a helping hand to Texas missions.

It has for years been my judgement that our missions and Sunday-school are handmaids, and ought to receive the united support of the denomination. Anti-Sunday-school men, if sounded, as a rule are anti-mission men. At the close of the war between the States, in 1865, evidences of demoralization were apparent among the youths at home, as well as among the returned soldiers. A desire to counteract these evil influences among the rising generation seized the minds of thoughtful Christians, and found earnest advocates among Sunday-school men.

At the meeting of the Union Association in the old town of Washington, in August, 1865, Elder S.I. Caldwell read a report on Sunday-schools, and in that report recommended the organization of a separate Sunday-school Convention.

The report was adopted, and the organization was formed at Independence in October of the same year, with Judge A.S. Broadus as first president. According to my best recollection, this body has not failed to meet every year since, assembling as far east as Tyler, as far north as Waco, and as far west as Gonzales. Sunday-schools and Workers' Institutes have been organized through the influences and under the auspices of this Convention in a large number of our District Associations, and interest and information have been imparted that have kept Baptists all over Texas abreast with any people in the Sunday-school work.

For about two years after the organization of the "Texas Baptist Sunday-school and Colportage Convention" Elder S.S. Cross was the general agent. After he retired, owing to some confusion in the body, but little was accomplished until the Convention met in Austin, in 1872. At that meeting Elder M.V. Smith, then pastor at Brenham, was chosen as agent, and during the fifteen months that he served succeeded in rallying the denomination and inspiring hope for the future. It remained however for Elder W.D. Powell to lead the denomination to the point of success assured in this department. He has now been in the field about five years, and but few men have performed a greater amount of effective labor in the same length of time.

He is a scholar— plain, child-like, senisble, practical and earnest. He chains the hearts of children to his own, and of course by this means gets the hearts of their parents and friends. He walks, he drives, in haste, and seldom finds a locomotive on iron rails that has speed enough for him; and yet his haste never amounts to confusion. But few men are equal to him as an organizer, and in that lies his great power. He has about him the elements of success, and has done a grand work.

During the past year the Convention has employed an additional Sunday-school missionary, in the person of Elder Kitt Williams. From all I have seen and heard, brethren Powell and Williams are indeed "co-workers," and 'fellow-helpers to the truth."

Elder Williams has been reared in Texas, and graduated at Waco University. As a temperance worker he has had few equals. He understands frontier life, and is well qualified for the work in which he is engaged. He is a preacher of ability — clear, logical and scriptural—with an address so easy and natural that the man for the time is forgotten and truth studied. These, as I understand, are the elements of true oratory. Long may these men and the Convention they represent live, and may the end of their labor be realized—a Baptist Sunday-school in every church in Texas. All who have had opportunity for observation must admit that a large amount of our present prosperity is due to the work of the "Texas Baptist Sunday-school and Colportage Convention."

I attended the meeting of this Convention at Tyler in July, 1873, and for the first time met W.E. Penn, of Jefferson, who was elected President of the Convention, and who filled the same position the following year in the city of Houston. He was at that time Superintendent of the Baptist Sunday-school in

Jefferson, and his wonderful success in that school was attracting the attention of the denomination.

Major Penn, as we all then called him, was born in Rutherford county, Tennessee, the 11th day of August, 1832. He was the son of Geo. D. and T. P. Penn. He was reared on a farm, and received his education in "the old field schools" of that day, except one year spent at Murfreesboro, Tennessee.

At the age of fifteen years he professed conversion, and was baptized. He selected the law for his profession and was admitted to the bar at the age of twenty-one, and spent twenty-two years in the courts of Tennessee and Texas. At the age of twenty-three years he was married to Miss Corrilla F. Sayle, the daughter of Dr. S.G. and Mrs. Olivia W. Sayle. He entered the Confederate army early in the war and served to the close, attaining the rank of Major. He saw much hard service, and spent the last two years of the war a prisoner on Johnson's Island. About the time of the surrender of General Lee and the Confederate forces east of the Mississippi Maj. Penn was exchanged, and reported for duty in the Trans-Mississippi department. Broken up by the war, and seeing what he took to be a better opportunity to recover his losses in Texas than in Tennessee, after the surrender of our western army he settled in the town of Jefferson and entered vigorously upon his profession in 1865. There he made friends, money and reputation for ten years.

Two years after I first met him he attended the meeting of a Sunday-school Institute in the same town of Tyler, in July, 1875. Elder James H. Stribbling was at the time pastor of the church at Tyler. He with other brethren were impressed with the belief that brother Penn ought to give up the law and devote himself to Sunday-school work for the Convention, and perhaps enter the ministry.

Thus impressed, the pastor compelled Major Penn to conduct a prayer-meeting, much against his will, the morning he and his party had appointed to return to Jefferson. At the close of that service he was urged and prevailed upon to stay and conduct a service at night. At that night service he made his first appeal to sinners, and a number came forward for prayer. As he was about to close the meeting Elder Stribbling stepped forward and asked the people, both saints and sinners, to rise, who were willing to join in an earnest protest against the return of Major Penn the following day to Jefferson. He who faced the musket and cannon in battle, and who had been hardened by prison life on Johnson's Island, trembled before that manifestation, and feared in the midst of a praying people to refuse to stay.

His wife and the party who came with him returned to Jefferson, his law business was laid aside, and he remained in Tyler, conducting two and three and sometimes four services a day, for five weeks. Many were drawn to church who seldom came, and all classes of people were reached. About one hundred professed conversion.

At the close of these services he returned home and entered his law office with fear and trembling, lest by this long absense the interests of his clients had suffered and his standing as an attorney had been ruined. He well knew that it was the disposition of people to turn from a lawyer who turns preacher, upon the same principle that rats desert a sinking ship. A careful examination of the letters received in his absence showed that his clients' interests had not suffered, and his faith in God at this trying moment assured him that if God intended to put him into the ministry he need not trouble about his reputation as a lawyer. As the condition of affairs pressed upon his mind, in imagination he was a man in the midst of ashes. The profession that had been his bride for twenty-two years, and his means of support, like a mist vefore the rising sun vanished. His well-furnished office, his library and home, where he had spent money on flowers and paintings and comforts, with an open hand—all, all seemed like a beautiful cloud melting away. In this state of mind, halting and troubled, he remained for weeks. In the meantime the way opened to dispose of the cases he had in court, some of them long standing. This was an additional evidence that Providence had been fitting him for the work of an evangelist in Texas.

He was at this time just 43 years of age, 6 feet high, and weighed 218 pounds. With black hair, the top of his head bald, with very dark hazel eyes underneath a very broad and prominent forehead, a round, full face, a mouth unusually large, with long, flowing beard, and a heavy bass voice. Any man who sits in his audience, and for the first time listens to that voice in song and earnest, importuante prayer, will expect to hear something in the sermon worth listening at. In this he will not be disappointed. If he expects flowers and rhetoric and poetry, he may be disappointed. But if he expects to be brought face to face with his God, told of the corruption of his heart, the absolute necessity of a change in his nature that can only be effected by the Spirit of God applying the blood of Jesus, and that a hell is beneath him of "fire and brimstone" that may be reached at any moment, while heaven invites the wanderer—if the listener expects to hear the plainest, most earnest and pointed talk of thirty minutes he ever heard on these and kindred themes, he will not be disappointed.

The news of the Tyler meeting spread rapidly over the State, and invitations came to visit other points. After much thought, prayer and correspondence with Elder F.M. Law, pastor of the church at Bryan, he was next induced to go there, and a meeting was conducted similar to the one at Tyler and with about like results. There I first met him as an evangelist, and have ever since been thankful for the meeting.

Returning from one of his day meetings, and sitting in Elder Law's office, he told me that a dark cloud was upon him. He said the meeting that day seemed to him a failure, and in a desponding state he told me of his misgivings relative to the work he was doing. He said it was evident that what means he had would in a short time be gone if he persisted in conducting these meetings, and that he was sorely troubled.

I had passed a similar ordeal many years before. I told him my experience, and brought before him the word of God. I told him that He who had redeemed his soul, protected him in battle and in prison, and was blessing him as an instrument in saving sinners, would in some way care for his actual wants. I told him, however, that he must not expect to get money as he had done, but that he who worked for Jesus and for souls must be content with food and raiment. He buried his face in his hands and sighed. I shall not forget the silence that followed. In a few minutes he decided a question that took me three years to settle. He raised his head, and looking me full in the face said: "That is a sensible view of the subject, and I will try it." He at once turned to his bible, and began his preparation for the night service. After this decision God gave him great liberty, and that night the Spirit of God was with the preacher and upon the people in great power.

He has now been in the field seven years, and out of 2,555 days he has only lost 150 days from travel, sickness and rest. All this time he has conducted from two to four services a day. He has conducted meetings with 70 churches, besides 25 camp meetings. He has held meetings in five States besides Texas, viz.: Louisiana, Alabama, Tennessee, Missouri and Kentucky, and has not avoided large cities, nor passed unheeded petitions for help in rural districts.

There have been about sixteen thousand professions of conversion under his ministry. About twenty-five hundred of these were at the time members of churches, and about five hundred were infidels and skeptics of different grades.

It is a remarkable fact that he has never been too hoarse to sing and preach.

He began this work weighing 218 pounds. Once he weighed 214 pounds, averages about 220, but now weighs 231 pounds.

The largest number of conversions in one meeting was 450, and 349 of that number were baptized. The largest number of conversions in a day was 100, and the largest number at one service was 60. The smallest number at any one place where he held a meeting was 33, and he was only there three days.

This uniform and wonderful success has puzzled many a brain besides mine. Since the days of the apostles I have not read of anything like it, continued so long. Please turn and read 2nd Cor., ii. 14.

In 1879, having been in this work for four years, and having tried brush arbors in summer for the thousands who came from near and from far to attend his meetings, he conceived the idea of raising means to purchase a large tent. This was done at an expense of over five hundred dollars, and he has been spreading this tent for the past few years in various localities in Texas, from June till November. His power was never fully known till he was equipped and fully organized to carry on his camp meetings in summer, regardless of weather. He

keeps an organist and carries an instrument with him. The following are the names of the organists respectively who have traveled with him: V.C. Hart, J.F. Parker, H.A. Somrell, H.L. Alison, J.M. Hunt and L.B. Shook. From one cause or another he was much troubled to keep an organist with him. After making this matter a subject of special prayer, brother Shook, the last named and the present organist, was converted in a camp meeting last year at Flatonia, and baptized at Willow Spring in Bell county by brother Penn.
county, last November, and my health was then so feeble that I could only attend a few services. As I have felt it my duty to put to record this lengthy notice of this remarkable man and his wonderful work, I must state that I am largely indebted for facts and impressions here given to pastors and brethren where he has held meetings; but I am under special obligations to Elder M.V. Smith, of Belton, who has been with brother Penn in his meetings more than any other pastor. He has held four meetings in Bell country alone, one with the church at Belton and three camp meetings in different localities. In these four meetings there have been 1,014 professions of conversion.

For four years and a few months over, after he began the work of an evangelist, he refused to be ordianed. Finding, however, that in occasional cases it was not convenient to get ordained preachers to do the baptizing, he consented, and was duly set apart to this work also by imposition of hands on December 4th, 1879.

When a man undertakes to give the secret of his power over the people he finds it a difficult matter. He who ascribes it to any one thing makes a great mistake. I have studied him, and his work, both by inquiring and observation. I record the following, and if any one can improve on it I will be under obligation:

1. He is a man of destiny, appointed by the Lord to do this very work.

2. He has more faith that God will answer prayer, and answer it at once, than any man I have ever seen.

3. He is a man of decision. As questions arise he settles them in his mind at once, as he did the question of his support raised at Bryan. Some men spend much of their time and energies wrestling with doubts. Brother Penn spends all his time at work, instead of parleying with Satan, and hence that positivism to which some object.

4. He is outspoken in all his convictions. He hates hypocrisy in every form and in every body. What he is, he is by the grace of God. On this line he has made his worst enemies. He is a Baptist, and is not ashamed of it, nor afraid to tell the reasons why.

5. He is wholly consecrated to this one work.

6. Last, but not least, he is a general. He is wonderfully gifted with common sense. He concentrates all minds on one thing, and puts everybody about him to work to accomplish that one thing.

I believe he is a man of God, and urge my brethren to rally around him as long as he is faithful.

The agitation among our people on the subject of education for the past ten years is an evidence of the appreciation Baptists have of the importance of mental training, under proper influences. Side by side our mission and educational interests have stood in all our past history, and such is likely to be the case in the future. Though there have been differences of opinion and sharp controversies, more than were desirable, the cause has gradually gone forward. Baylor University, with the same president who was in charge ten years ago, still survives, and yearly reports progress. Only a short time after I finished my former writing, R.E.B. Baylor, for whom this institution was named, died, at a ripe old age, and by his request was buried near the walls of the building he loved in life and remembered in death. I saw him only a few days before he died. He was calm and resigned, and when wading into the waters of death showed himself to be the same mighty man in Israel that he was in the vigor of manhood.

Competent judges decide that Baylor Female College was never in so prosperous and hopeful a condition as it now is, under the management of Elder J.H. Luther.

Waco University, with the same president, continues and increases in prosperity. Since my last writing Elder R.B. Burleson has passed away. His name has been associated with Baptist educational interests almost from the beginning. When he died Texas lost one of the purest Christians, ripest scholars, best educators and most reliable gospel preachers it was ever my fortune to know.

Schools in which we feel a deep and abiding interest have sprung up, and are springing up, all over our State, in which the children of our brethren are receiving good training, and where young preachers who have not the means to go abroad can be educated at home, and continue their ministry among our churches.

I cannot give the long list of names, but will briefly refer to two, which have been under my observation. In the new town of Kyle, within three miles of my present home, in a healthy locality at the foot of the mountains, is a splendid school building that cost between three and four thousand dollars. In this building is a school of over one hundred students, with Elder T.H. Storts as president. He presided over a female school in Kentucky for fourteen years in succession, and by this record shows that he is no wandering comet. He is a fine scholar, with an accomplished family, a good preacher, and offers advantages long needed in Hays county.

In the same county of Hays, only twenty miles north of Kyle, right among the mountains, is a large stone building at Dripping Springs, with a school of over sixty students. Here is a community organized with reference to health and education. Elder Wm. Jordan, from Indiana, is president of this school. As a disciplinarian I have never observed his superior. His wife teaches music. He is a ripe scholar and an able preacher. He has three young men studying for the ministry, and others are coming. From this community, noted for good health and good morals (where no whiskey can be sold), and from this excellent school, I expect educated men and women to go forth to bless Texas and the world when I am gone.

Among the records I am called upon to make with joy, duty calls me to pen at the close death of three of the noblest men I ever knew, and who have for so many years been my true and tried friends. The year 1881 stands without a parallel in my recollection for sad events throughout the civilized world. Who ever read of the death of so many good men, great men and representative men, in one year? Beneath the power of the spirit of communism, nihilism and moboracy, that conceives and organizes in secret and at night, that brought last year the Czar of Russia to his grave, ended the life of our President, that induced strikes among the employees of railroads and manufacturing companies, and that has dragged prisoners from the jail and hanged them in the night, the nations of earth have trembled. Tortured almost beyond endurance by the news of these sad events, flashed over the wires last year in rapid succession, we received among other things notice of the death of Jonas Johnston in April, J.W.D. Creath in July, and H.L. Graves in October.

Sad to us all was the announcement that Elder Jonas Johnston was dead, it rejoices my heart to know that he preserved till death the same character I recorded of him ten years ago. I feel it my duty here to state that after I had written originally, by request of brethren privately made and by request of the Texas Baptist State Convention, that I found myself without money to publish what I had written. When brother Johnston was apprised of the situation, he promptly agreed to furnish half the money required for one thousand copies, and brother C.R. Breedlove, of Brenham, furnished the remainder. When the money was returned they refused to receive any interest. When the stereotype plates were burned they furnished the money again without interest. If the work has done or shall do any good, these brethren share in it largely. As a financier, brother Johnston knew no superior among our ministry. A few years before he died he bought an interest in the "Texas Baptist Herald," and was the business manager till he died. Looking after finances did not interfere with his preaching. The best sermons I ever heard him preach were delivered while traveling in the interests of this paper. He suffered long and died at home. When asked by his pastor relative to his hope, he simply replied, with a smile, "That has been settled long ago."

Elder J.W.D. Creath died near Cameron, at the house of Elder T.M. Anderson. No man died with the harness nearer all on and better adjusted than brother Creath. He continued traveling for the State Convention till he died. He seemed as one craving death either in the pulpit or in the saddle of a missionary; and if this were his desire, he was not far from either when death overtook him. No man ever did more for Texas than he did. It is sad to think of his dying from home, but brother Creath had a home in almost every community, and in the house of every good man in Texas. His was a grand life and a glorious death. He was arranging appointments ahead when death knocked and call him by name. Recognizing the messenger, he said to friends around as he was passing away, "Sweet rivers of redeeming love are before my eyes."

I recorded the fact that in 1846 there came to Texas on the same ship four able ministers of the gospel, Elders Taliaferro, Creath, Graves and Chandler. Bro. Taliaferro died several years ago. Last July brother Creath died, and in October we received news of the death of brother H.L. Graves, leaving as the only survivor of that cargo of preachers Elder P.B. Chandler, who still lives near Gatesville, in Coreyelle county, and is pastor of churches around him.

The last few years that Elder H.L. Graves lived were spent in an active ministry in the south-west. He attended the meeting of the Texas Baptist State Convention and of the Educatoin Society, of which he was president, at their last session in Galveston, the first of October last, and went to his home in Brenham apparently in good health. He was taken sick at once, and after much acute suffering, which he bore with great patience, he passed away. I have never known a better presiding officer and more amiable Christian gentleman than Henry L. Graves.

But very few now remain of those who took part in our early organizations. While announcing the death of these leaders among Baptists with sorrow, I am but the more forcibly reminded of the fact that I am left almost alone, since the recent record has been made of the death of Elder Robert Alexander, at Chappell Hill, who was the first missionary sent to Texas by the Methodist church. We labored on the same field in an early day. He was uniformly mild and courteous, a good preacher and a useful man. He was much younger and stronger than I was when first we met, and yet he with about all my early associates are gone, while I linger, waiting for the command of my Lord to join them on a field of triumph.

CHAPTER XXXIV.

A CONTRAST: 1872—1882.

BEFORE reading this chapter, turn again and read chapter xxxii.

In 1872 the population of Texas was about 1,000,000. Now the population is 2,000,000.

Then there were in Texas 536 post-offices. Now there are 1,302.

Then there were 109 newspapers. Now there are between 400 and 500.

There were ten years ago 1,000 miles of railroads. Now there are 4,389.

The public school lands of Texas amount to 40,000,000 acres, worth about $100,000,000. The University of Texas, soon to be built, has $500,000 with which to erect buildings, and a permanent endowment of $2,000,000 or $3,000,000.

There is in our state treasury a cash balance of $800,000.

In 1872 Baptists numbered 50,000. Now they number 80,000 whites and 40,000 colored; total, 120,000.

We have 1,500 white churches and 600 colored; total, 2,100.

There are 1,300 white Baptist preachers and 500 colored; total, 1,800.

We have 60 white Associations and 25 colored; total, 85.

We have 5 white general bodies and 2 colored; total, 7.

We have 3 Baptist colleges and 4 Baptist newspapers.

Should I live to see January 17, 1883, I will be 80 years of age. On the 9th day of May last, 60 years ago, I professed a hope in Christ, was baptized on the 22nd day of the same month, and have been preaching ever since. Forty-six years of this time I have spent in Texas. Under the providence of God I have lived to see one preacher multiplied into 1,800, one church into 2,100, and less than 50 Baptists into 120,000. Whether I shall be permitted to take any further part in work and records of work has with me ceased to be a matter of much anxiety. I rejoice over what I have seen, and am still hopeful relative to the future.

The End

INDEX

by BEN ROGERS

CONVENTIONS

INSTITUTIONS

PERSONALITIES

A

PLACES

TOPICS